Timing Verification of Application-Specific Integrated Circuits (ASICs)

Timing Verification of Application-Specific Integrated Circuits (ASICs)

Farzad Nekoogar
Lecturer, Department of Applied Science,
University of California at Davis

Prentice Hall PTR
Upper Saddle River, NJ 07458
http://www.phptr.com

Library of Congress Cataloging-in-Publication Data

```
Nekoogar, Farzad.
     Timing verification of application-specific integrated circuits
   (ASICs) / Farzad Nekoogar.
       p.  cm.
     Includes bibliographical references and index.
     ISBN 0-13-794348-2 (case)
     1. Application specific integrated circuits—Design and
   construction.   2. Integrated circuits—Verification.   I. Title.
   TK7874.6.N45.  1999
   621.39'5—dc21                                      99-33309
                                                          CIP
```

Editorial/production supervision: BooksCraft, Inc., Indianapolis, IN
Acquisitions Editor: Bernard Goodwin
Editorial Assistant: Diane Spina
Marketing Manager: Lisa Konzelmann
Manufacturing Manager: Alan Fischer
Cover Design Director: Jerry Votta
Cover Designer: Talar Agasyan
Project Coordinator: Anne Trowbridge

The publisher offers discounts on this book when ordered in bulk quantities. For more information contact:
Corporate Sales Department
Phone: 800-382-3419 Fax: 201-236-7141
E-mail: corpsales@prenhall.com

Or write:
Prentice Hall PTR
Corp. Sales Dept.
One Lake Street
Upper Saddle River, NJ 07458

Printed in the United States of America
10 9 8 7 6 5 4 3 2 1

ISBN 0-13-794348-2 (case)

Prentice-Hall International (UK) Limited, *London*
Prentice-Hall of Australia Pty. Limited, *Sydney*
Prentice-Hall Canada Inc., *Toronto*
Prentice-Hall Hispanoamericana, S.A., *Mexico*
Prentice-Hall of India Private Limited, *New Delhi*
Prentice-Hall of Japan, Inc., *Tokyo*
Prentice-Hall (Singapore) Pte. Ltd., *Singapore*
Editora Prentice-Hall do Brasil, Ltda., *Rio de Janeiro*

I dedicate this book

to the memory of my grandparents

—Farzad

Contents

List of Figures

List of Tables

Preface

This book describes the theory and applications of timing verification of application-specific integrated circuits (ASICs). Timing verification is a relatively new concept, which is why most books on digital systems do not cover the issue. This book lays out the fundamental principles of effective timing verification, and it makes good use of the examples that reflect the current issues facing logic designers.

The following items characterize this book:

- ☞ Timing verification as opposed to functional verification is the primary focus. (Functional simulation has been adequately covered in other books.)
- ☞ Principles and techniques as opposed to specific tools are emphasized. Once designers understand the underlying principles of timing analysis, they can apply them with various timing tools.
- ☞ Design flow for deep-submicron ASICs and FPGA designs are fully covered.
- ☞ Numerous design examples and HDL codes to illustrate the concepts discussed in the book are provided.

This book is to be used for self-study by practicing engineers. Design and verification engineers who are working with ASICs and FPGAs will find the book very useful. Upper-level undergraduate and graduate students in electrical engineering can use it as a reference book in design courses in timing analysis and related topics.

The material covered in this book requires some understanding of the Electronic Design Automation (EDA) tools and an initial course in logic design.

The book is organized into two parts:

Part I (chapters 1 and 2) introduces the fundamental concepts involved in timing verification. Including clock definitions, multicycle paths, false paths, and phase-locked loops.

Part II (chapters 3 and 4) covers specific timing issues related to ASICs and FPGAs, respectively.

Chapter 1 gives an overview of timing verification and static timing analysis. It contrasts timing verification with functional verification. Typical goals of timing verification in digital systems are presented. This chapter ends with an example of interface timing analysis.

Chapter 2 introduces the concepts of timing analysis with design examples. It specifically discusses such clocking methods as gated clocks, multifrequency clocks, and multiphase clocks. It introduces the concepts of multicycle paths, false paths, and timing constraints (such as setup, hold, recovery, and pulse width).

Chapter 3 discusses the deep submicron ASIC design flow and application of timing analysis in the design process. It includes discussion of prelayout and postlayout timing verification. The chapter also discusses behavioral and structural RTL coding for timing, synthesis and timing constraint, and the ASIC sign-off checklist. We make the concepts concrete with numerous examples.

Chapter 4 discusses timing concepts in programmable logic-based designs. It covers design flow, timing parameters, timing analysis, and HDL synthesis and software development systems. We present the most commonly used programmable logic devices (Actel, Altera, and Xilinx) and associated timing issues.

Appendices A, B, and C discuss the EDA timing tools of PrimeTime, Pearl, and TimingDesigner respectively.

Appendix D covers some concepts of transistor-level timing verification.

ACKNOWLEDGMENTS

I would like to thank the following individuals from Intrinsix Corp. for their contributions to this book:

- Lawrence Letham, for authoring chapter 3 and contributing to other topics.
- Paul Brown, for authoring chapter 4.
- Faranak Nekoogar, for her knowledge of PrimeTime and her help on chapter 2 and appendix A.
- Gene Petili, for his contribution on the subject of transistor-level timing verification.
- Peter J. Militello, for his help in the area of phase-locked loops.
- Tomislav Ilic, for his help on chapter 2.
- Mark Beal, for the help he provided to this project.

In addition, I'd like to thank the following people and companies:

- The staff of Prentice Hall, especially Bernard Goodwin, for his support of this project.
- Wilbur Luo, director of field applications, Chronology Corporation.
- Terry Strickland, director of marketing, Chronology Corporation.
- The staff of BooksCraft, Inc., for their help in producing the book.

—Farzad Nekoogar

Timing Verification of Application-Specific Integrated Circuits (ASICs)

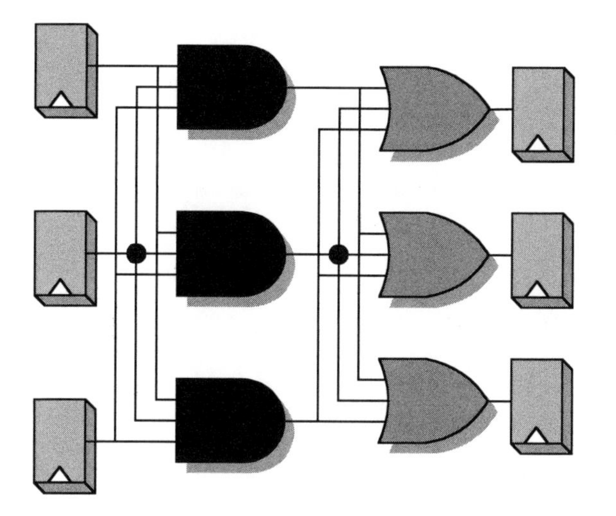

1 Introduction to Timing Verification

1.1 INTRODUCTION

In this book we introduce the concept of timing verification of ASICs. The fundamental concepts can be applied to ASIC, field programmable gate array (FPGA), and system-on-a-chip designs. It should be noted that most of the material presented in this book is electronic design automation (EDA)-tool independent. However, any mention of specific commands for Static Timing Verification (STA) is based on the Synopsys PrimeTime static timing analyzer. Also, any mention of a specific synthesis tool or command refers to Synopsys Design Compiler and its related modules.

Overview of some current EDA tools that are used for timing verification can be found in appendices A, B, and C. Also, Verilog-HDL is used throughout the book for hardware description language (HDL) coding examples.

In this chapter we cover the intrinsic versus extrinsic and static versus dynamic timing verification. We present the concept of path delay, which is the fundamental concept in STA. We also discuss timing interface, and an example of interfacing an Intel i960 processor to an EMS EDRAM through an FPGA that acts as the memory controller is presented.

1.2 Overview of Timing Verification

Figure 1.1 shows a typical ASIC design flow. The flow can be divided into the following major parts: design entry, design implementation, design verification, and integrated circuit (IC) production. This is the conventional design flow, and timing verification is part of design verification.

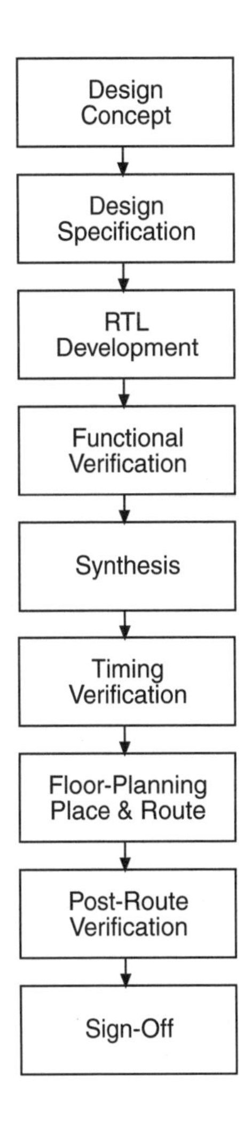

Fig. 1.1 Typical ASIC Design Flow

Functional simulation is the first necessary step after a design is completed. Functional simulation tests the entire design for its functional requirements. Once the functional requirement is checked, the timing verification determines if the design meets its timing requirements and if the delays can be optimized for better circuit performance. Dynamic timing simulation and static timing analysis are both essential methods for verifying the timing characteristics of a design.

Dynamic timing simulation can be used for timing analysis of asynchronous designs as well as synchronous designs. Dynamic timing simulation requires comprehensive input vectors to check the timing characteristics of critical paths in a design. Dynamic simulation can verify the functionality of a design as well as the timing requirements.

Before the advent of static timing analyzers, the only way to verify that a circuit worked at speed was to do exhaustive simulations on back-annotated gates. The problem with verification through simulation is that gate-level simulations are slow and require significantly more time to run the same test vectors used during logic development. Completeness of the simulations is also an issue. Simulation cannot prove that a design is free from errors, just that there are no problems when tested a specific way. Slow gate-level simulations provide poor coverage. STA was developed to offer an alternative to gate-level simulations before fabrication. In practice, STA is used to verify the timing of the entire device while a few critical vectors are run on the gate-level model. Static timing verification checks every path in a design for timing violations without checking the functionality of the design. Therefore, the functional and timing verification efforts can be done in parallel. This method is significantly faster than dynamic timing simulation because no test vectors need to be generated for timing verification. Static timing verification has sophisticated analysis features such as false path detection and elimination, and minimum/maximum analysis to generate comprehensive reports. However, static timing verification can be used to verify only the timing requirements of synchronous circuits.

STA verifies the delays within a circuit. It is capable of verifying every path and can detect serious problems like glitches on the clock, violated setup and hold times, slow paths, and excessive clock skew. It offers verification times that are a fraction of what it takes to do gate-level, back-annotated simulations. Although STA is practically limited to synchronous circuits, its ability to provide comprehensive, quick verification has made it the final sign-off tool with many vendors.

The other emerging design methodology that is becoming popular with designers is spiral design flow. Figure 1.2 shows this flow. Here the designers work simultaneously on each phase of the design until the design is gradually completed. This type of methodology is used mostly in large system-on-a-chip designs.

1.2.1 Intrinsic vs. Extrinsic Delay

If you consider a logic gate to be a black box with input and output pins, the intrinsic delay would be the delay internal to the black box

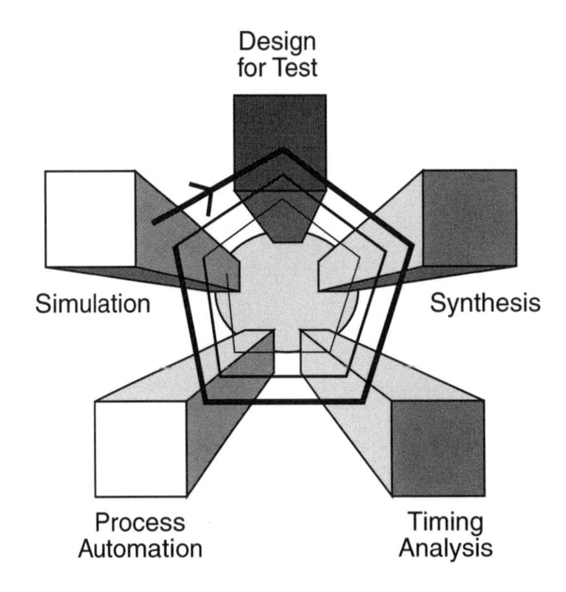

Fig. 1.2 Spiral Design Flow

(pin to pin) and the extrinsic delay would be the delay between black boxes (point to point). Digital design tools then define the black box and the underlying functionality using Boolean equations and timing arcs for the intrinsic delays. The output drive strength (resistance), input capacitance, and wire load models are used to calculate the extrinsic delay between logic gates. Vendors often specify the drive strength as ns/pf or ns/std_load, which both can be units reduced into ohms. Timing analysis then becomes a case of solving the simple linear equation $y = mx + b$ where y is the total delay, m is the output resistance, x is the load capacitance (measured as pf or std_loads), and b is the intrinsic delay.

Digital tools get a 500+ speed improvement over analog simulators due in part to the precalculation of the circuit delays. The intrinsic delays are usually fixed in the model and the extrinsic delays are calculated on a node by node basis once the analysis is started. Separate delay triplets are usually calculated for rising and falling edges on both the input (intrinsic) and output (extrinsic). Once calculated, the delays are held constant for the remainder of the analysis.

The digital tools have been increasing the number of factors they include in the analysis. However, the nature of static timing analysis depends on the basic assumption that all delays can be calculated at the beginning of the analysis. This implies that the delays are not data dependent. This is not a valid approximation for deep-submicron designs, which exhibit state-dependent characteristics that do not lend themselves to static calculation.

Some dominant sources of delay are:

- ☞ Input capacitance of the logic gate is a function of output state, output loads, and input slew rate.
- ☞ Multiple inputs switching can have additive and subtractive effects, which can increase and decrease intrinsic timing arcs as well as output slew rates.
- ☞ Capacitance of the wires is frequency, slope, dependent. Given the slope's data dependency, the wire load can exhibit additional data dependency.

Other slope-dependent effects that are often neglected include:

- ☞ Internal timing arcs are a function of input slew rates (i.e., intrinsic delay is a function of extrinsic).
- ☞ Output resistance, and, therefore, output slew rate, is a function of input slew rate on each input (i.e., extrinsic is a function of extrinsic).
- ☞ Wire loads exhibit RLC tree characteristics instead of simpler lumped RC.
- ☞ Transmission line effects cause receivers closer to the driver to see an input sooner and with a steeper slope than receivers further from the driver. Single logical signals need to be broken into multiple sequential signal paths.

These effects existed in older 1 micron processes; however, they were insignificant compared to the static portion of the intrinsic delay. With intrinsic gate delays of deep-submicron processes measured in pico seconds, these once third-order effects (sometimes referred to as routing delays) are often the dominant source of delay.

Since digital tools do not fully account for deep-submicron effects, library designers resort to overly restrictive design rules to ensure that their designs will function. Often, one or more of these effects is ignored and will cause the real silicon not to meet timing verification even though the STA model did. Analog simulations are also often used for final timing verification of the critical paths identified by the STA. Most STAs will output spice netlists of the critical path to aid in this analysis. In extreme cases, designers may find it necessary to rely on more accurate, although much slower, analog tools to verify and/or characterize entire black-box subsystems. These accurate models are then used by the STA in place of the individual logic gates for the larger system timing verification. Appendix D covers more on transistor-level timing verification including some of the commercial tools available to perform this task.

1.2.2 Path Delay

The fundamental aim of STA is to measure the delay of every path. The value for gate delays comes from the vendor library. Interconnect delays are either estimated during synthesis or extracted after place and route. The circuit shown in Figure 1.3 illustrates the basics of STA. There are several paths from the input terminals to the output. Each path traverses interconnect and goes through logic gates. The propagation delay through each gate is shown next to the gate as a circled number while each net has an interconnect delay of one. The delay through each gate and along each net is totaled to get a path's delay. The paths shown in Figure 1.3 and their respective delays are listed in Table 1.1.

The path with the maximum delay is a-out via c1-c2-c4-c5-c6. It takes 26 time units for the output to settle to its final value when determined by input a. The minimum delay is from input e to out with a delay of only four time units. After analysis, the analyzer compares the times to the minimum and maximum specified by the designer for correct operation. The paths that do not meet specification are reported as problems. The circuit shown in Figure 1.3 is asynchronous, combinatorial logic. It must be mentioned that STA,

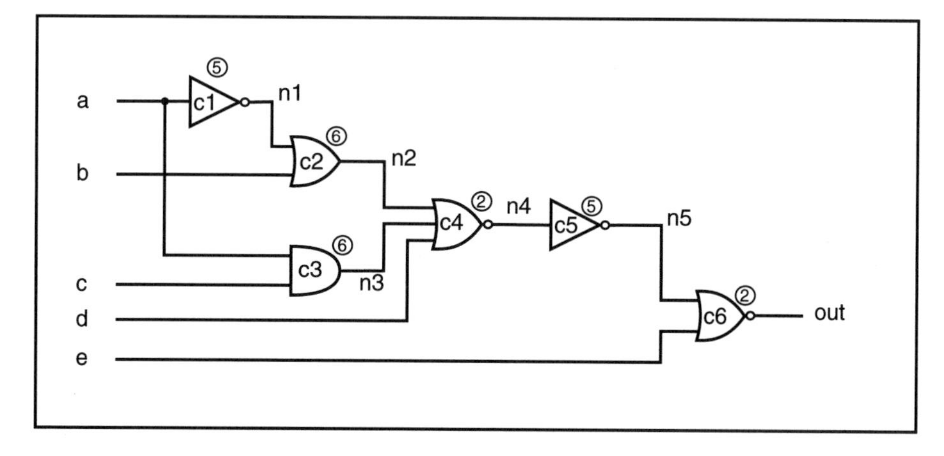

Fig. 1.3 STA Measures Path Delays Through a Circuit

Table 1.1 Delay Paths of Figure 1.3

Path	Signal Route											Delay
a – out	a	c1	n1	c2	n2	c4	n4	c5	n5	c6	out	26
a – out	a	c3	n3	c4	n4	c5	n5	c6	out			20
b – out	b	c2	n2	c4	n4	c5	n5	c6	out			20
c – out	c	c3	n3	c4	n4	c5	n5	c6	out			20
d – out	d	c4	n4	c5	n5	c6	out					13
e – out	e	c6	out									4

just like synthesis, is best suited for synchronous designs. It is possible to do some limited analysis on asynchronous paths, but functionality is not guaranteed. Asynchronous circuits still need to be verified via simulation.

A synchronous circuit is shown in Figure 1.4. Its minimum and maximum times are calculated much the same as for the previous circuit except the flip-flop propagation delay and setup times become part of the equation. The clock period, as defined by the designer, fixes the time allowed for propagation between each register. On the active edge of the clock, data propagates through the

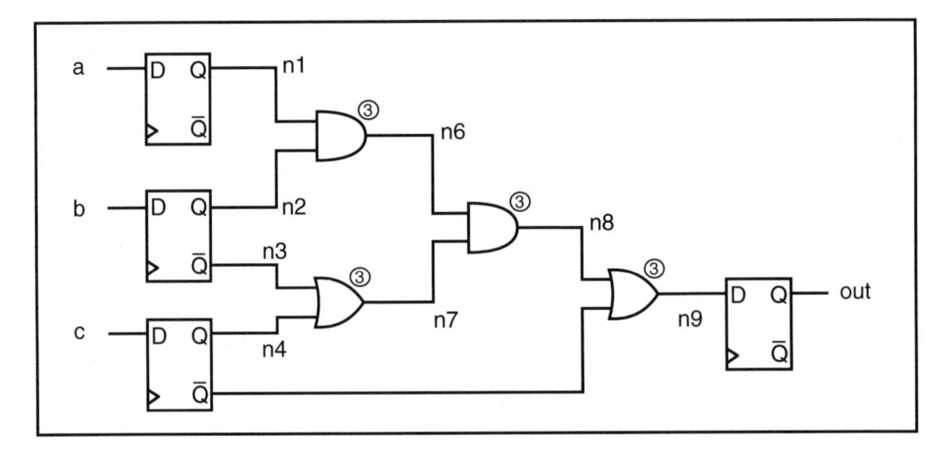

Fig. 1.4 STA Is Best Suited for Use on Synchronous Circuits

flip-flop gates and interconnect toward the next flip-flop. The data must arrive Tsetup before reaching the next active edge of the clock. If the clock period is 10ns and the data takes 15ns to propagate from the first flip-flop to the second, the data is obviously late by a wide margin. However, if the data arrives at the next flip-flop right at 10ns, it is still late because it has violated the flip-flop's setup time. The amount of time left for propagation between flip-flops as shown in Figure 1.5 is

$$\text{Tpropagation} = \text{Tperiod} - \text{Tsetup} \qquad \textbf{(Eq. 1.1)}$$

Table 1.2 details the path delays of Figure 1.4 as measured by STA. The propagation delay through the source flip-flop is added to the gate and net delays. The symbol Tq stands for the propagation delay from clock to q, and Tqb from clock to q bar. The setup time, Ts, is added to the other delays to arrive at the minimum frequency of operation. Once more, the gate delays are circled and each net has a delay of one time unit.

The examples of calculating path delays given above are simplistic. The actual is rather complex. Some approaches include techniques with names such as transition delay, floating delay, and timed Boolean functions [LB94]. The actual algorithms that calcu-

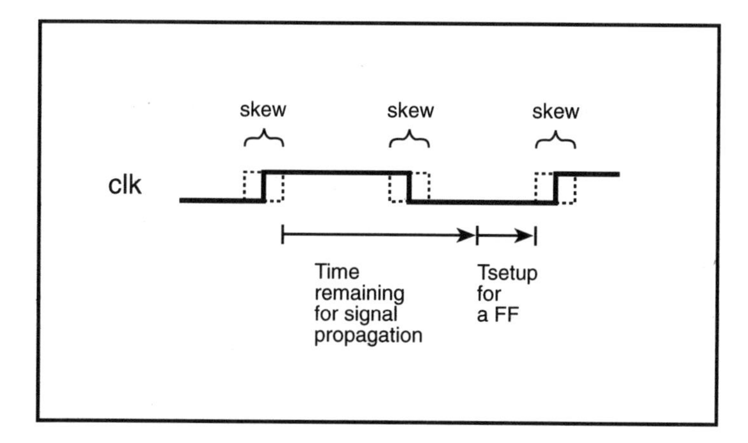

Fig. 1.5 Time Available for Propagation Between Two Flip-Flops Depends on Clock Skew and Flip-Flop Setup Time

Table 1.2 Paths Measured in STA

Path	Signal Route									Delay
a – n9	Tq	n1	c1	n6	c3	n8	c4	n9	Ts	13 + Tdq + Ts
b – n9	Tqb	n2	c1	n6	c3	n8	c4	n9	Ts	13 + Tqb + Ts
b – n9	Tq	n3	c2	n7	c3	n8	c4	n9	Ts	13 + Td + Ts
c – n9	Tq	n4	c2	n7	c3	n8	c4	n9	Ts	13 + Tq + Ts
c – n9	Tqb	n5	c4	n9	Ts	n8	c4	n9	Ts	13 + Tdq + Ts

late delay are beyond the scope of this book; however, some understanding of how an analyzer looks at a circuit is necessary to understand false paths.

A static timing analyzer must understand more than just adding up delays from tables. It needs to have some understanding of the circuit's logic. The truth table given in Table 1.3 analyzes the values of the input signals to the circuit shown in Figure 1.3 and identifies the path that determines the output value.

Table 1.3 Truth Table for Circuit in Figure 1.3

	Input Signals					
Line	**a**	**b**	**c**	**d**	**e**	**Dominant Path**
1	x	x	x	x	1	e-out
2	x	x	x	1	0	d-out
3	0	x	x	0	0	a-c1-c2-c4-c5-c6
4	1	0	0	0	0	a-c1-c2-c4-c5-c6 or b-c2-c4-c5-c6 or c-c3-c4-c5-c6 depending on transition
5	1	0	1	0	0	c-c3-c4-c5-c6
6	1	1	0	0	0	b-c2-c4-c5-c6
7	1	1	1	0	0	b-c2-c4-c5-c6 or c-c3-c4-c5-c6 slowest one

Table 1.3 reveals how the input signals determine which path through the circuit drives the output. Whenever the input signal e is one, the values of the other input signals are irrelevant because the one on input e forces the output of the NOR gate to zero. Essentially the same situation occurs on line 2 of Table 1.3. The value on input d determines the output value. Line 3 shows the conditions where the output depends on input a. Line 4 is a more interesting case because the path to the output depends on the order of input signal transitions. If the input signals are 00000 then become 10000, the transition goes through path c1-c2-c4-c5-c6. The transition 11000 to 10000 takes the c2-c4-c5-c6 path while the 10100 to 10000 transition goes through the c3-c4-c5-c6 path. In this circuit, the delays from b-out and c-out are the same, but gate and interconnect delays could be different and change the result. The inputs of lines 5 and 6 activate logic that forces the paths c-out and b-out respectively to determine the output value. The slowest path between b-out and c-out determines the delay in line 7 only if the transition is from 10000 to 11100. The delay path is already set and possibly settled with transitions from 10100 to 11100 (c-out) and 11000 to 11100 (b-out).

As a result of understanding the circuit's response to input stimuli, the timing analyzer knows that every path is capable of producing a response at the output, so it must consider all paths when determining delay. In this specific case, the longest path is from a-out through c1-c2-c4-c5-c6 and it is traversed during two different transitions: lines 3 and 4.

In chapter 2 we will consider false paths. False paths are logic paths that are not synthesized because they are functionally blocked. These paths are recognized by static timing analyzers as unconstrained paths. One example of false paths is the clocks that are not harmonically related to each other.

1.3 INTERFACE TIMING ANALYSIS

Some of the most common timing problems for ASIC, FPGA, or even board-level designs are related to interface timing between different components in a design. The following are some causes for some timing problems:

- ☞ Incorrect interface specification to the surrounding circuitry or between modules
- ☞ Timing problems caused by variations in components
- ☞ Oversimplified simulation models
- ☞ Incomplete test suites

Today's simulators and static timing analyzers cannot help you to produce timing or interface specifications. Specifically, these tools cannot answer design questions such as:

- ☞ What if a wait state is inserted?
- ☞ What if a faster part is used?
- ☞ How fast must the design run?

Timing diagrams are the best way to visualize potential timing problems early in the design process. They can be used to review performance requirements and communicate those requirements to all the designers on the project. A timing diagram represents not only the signal states over time, but also includes propagation delays, constraints, guarantees, and explanatory text annotations.

One EDA tool that helps you with generating timing diagrams is Chronology's TimingDesigner. It can be used to author the interface specification as well as perform worst-case timing analysis. Before we present an example, let's look at a generic logic design methodology.

1. **1.** Create timing diagrams (e.g., interface specifications) for the various bus cycles and operations.
2. **2.** Analyze the interfaces (e.g., worst-case timing analysis on the bus cycles).
3. **3.** Create bus functional models and your testbench (e.g., for dynamic simulation).
4. **4.** Create RTL (and test it in the testbench).
5. **5.** Synthesize (and test it in the testbench).

6. Perform gate-level and path-based static timing analysis.
7. Reanalyze the bus cycles using STA numbers (step 2).

Many of these steps outlined above have EDA tools to automate the process. For example, Chronology's TimingDesigner and QuickBench, both testbench generation products, can help with steps 1, 2, 3, and 7. The static-timing-analysis tools mentioned in step 6 would be complementary and used to provide more accurate data when the interfaces are reanalyzed. That is, the classic static-timing-analysis tools will calculate the path delays across the block or chip and these numbers can be plugged into the timing diagram to see if the cycle will work properly. Since static-timing-analysis tools have no concept of a bus cycle, using STA in conjunction with TimingDesigner, you not only analyze the paths but you analyze the bus cycle.

By analyzing the bus cycle, you can answer the questions: Do I need another wait state? Do I need a faster part? What if I repartition and change the specification and timing?

Example 1.1 (Courtesy of Chronology Corporation)

Suppose we have the task of interfacing an Intel i960 processor to an EMS EDRAM through an FPGA that acts as the memory controller. The question is: Will it work? And, if not, what are the options to make it work? Figure 1.6 shows the block diagram for our design.

The Intel i960 processor allows for burst operations and has four speed grades: 16, 25, 33, and 40MHz. The Intel iFX780 FPGA has a 9ns clock (clk) to output delay (synchronous path) and a 17ns clk to output delay (asynchronous path). The FPGA will perform address multiplexing (or muxing), decoding, refreshing, hit-and-miss comparing. The EDRAM has two speed grades—12 and 15ns.

For this example, we will do worst-case timing analysis on a four-word burst-read miss cycle to make sure the parts as specified above will work together properly. We will also optimize for cost and speed by choosing the slowest part that will still work.

There are four basic steps in creating and analyzing this design:

Example 1.1 (Courtesy of Chronology Corporation) (Continued)

1. Determine the bus cycles you are interested in analyzing.
2. Create, or obtain from the semiconductor vendor or Chronology, the diagrams for the bus cycles.
3. Confirm that diagrams have the bus cycle for the CPU connected to the bus cycle of the EDRAM.
4. Do the *what-if* timing analysis.

Figure 1.7 shows the merged read diagram of the CPU plus the EDRAM. The delays through the FPGA have been used. For example, the W/R# line of the CPU drives the W/R line of the EDRAM through the FPGA. W/R# is sampled by the FPGA on the second rising clock edge and there is a delay of 9ns (i.e., FPGA clock to output synchronous delay) through the FPGA to W/R.

Now we can perform the timing analysis. The analysis is true, worst-case analysis. Both the minimum and maximum delays will be accounted for along with the clock frequency and jitter. Any constraints that are violated will show up in red along with the margin.

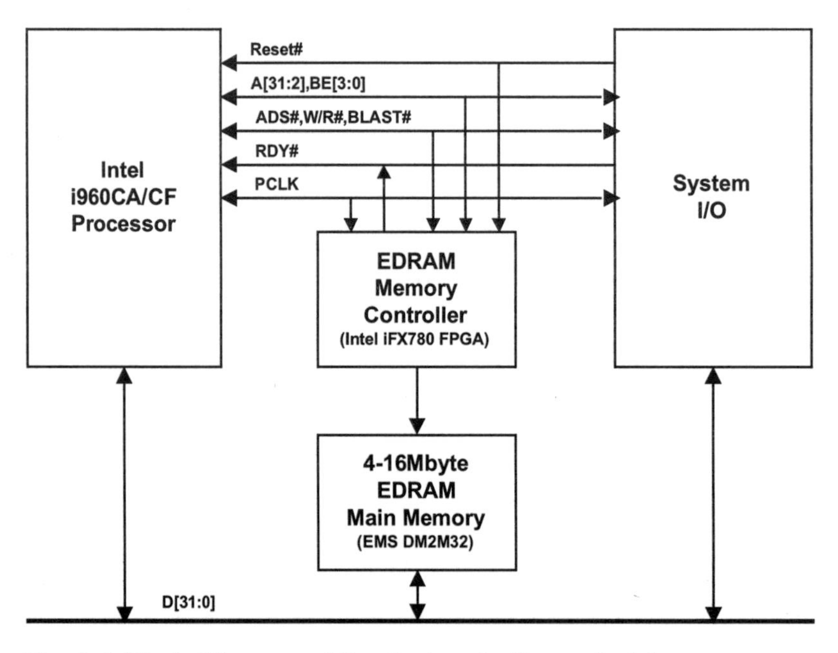

Fig. 1.6 Block-Diagram of the System in Example 1.1

In this example, we are using a 40MHz i960. It turns out that this configuration would be too fast and we would violate data setup times. (An example of a 2ns violation is circled in Figure 1.7.) We also note that we have incurred a wait state (shown in the diagram as an extra clock cycle to meet the speed of the EDRAM). With a spreadsheet (not shown) that is dynamically linked to the diagram we can do the what-if analysis. By simply swapping different speed parameters for the i960 and the EDRAM, we find that we can use a 16MHz processor with a 15ns EDRAM. If we had wanted to use the faster processor, we would need to use a 12ns EDRAM and have a 0ns margin.

In conclusion, we see that this technique of interface timing analysis can be performed at the beginning of the design cycle when the architecture and interface specification are still being formed. It

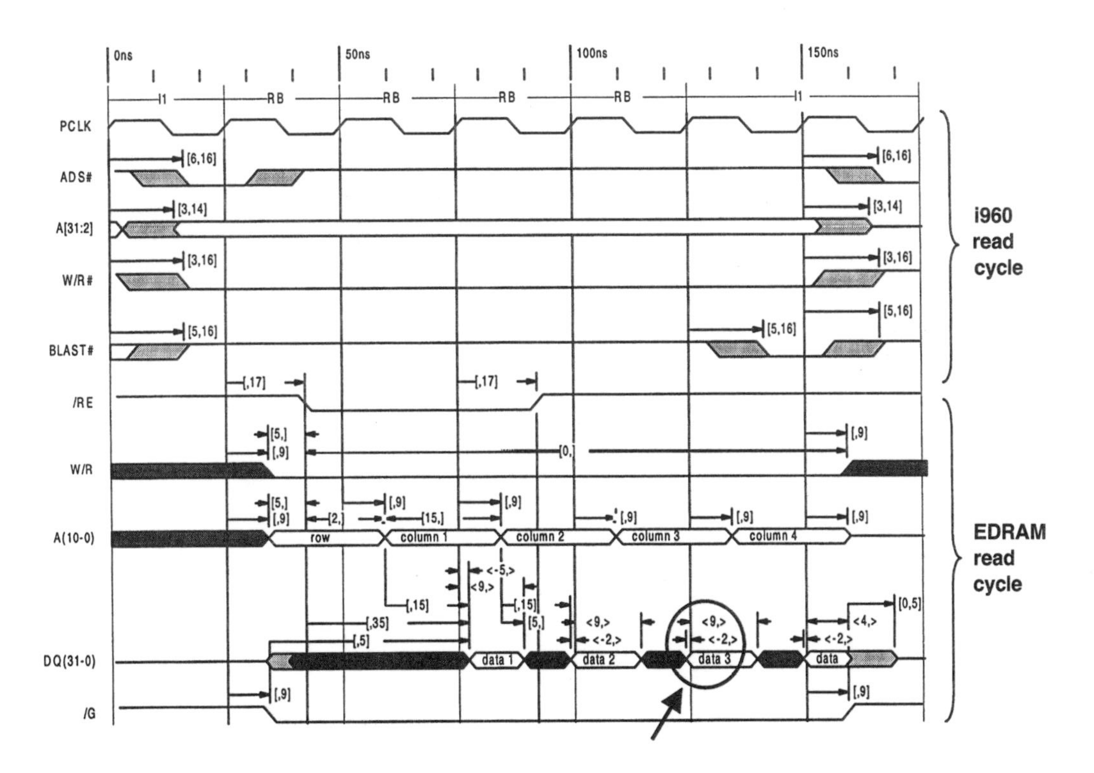

Fig. 1.7 Merged Read Diagram of the CPU Plus the EDRAM

can also be used when the specifications have changed or at the end of the design cycle to verify the final timing. By automating the worst-case timing analysis in context of the bus cycles, we have a simple method to do what-if analysis and increase the confidence that our interfaces will work correctly.

CHAPTER 2

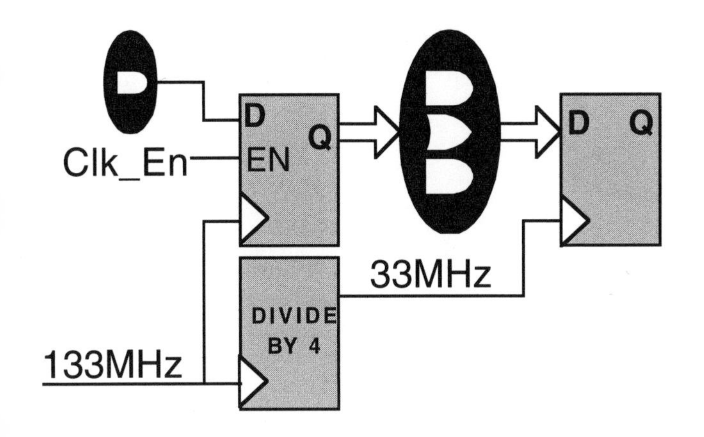

Elements of Timing Verification

2.1 Introduction

In this chapter we introduce and define some basic concepts related to synchronous devices. Clock definitions such as gated clocks, clock skews, multiple clock groups, multifrequency clocks, and multiphase clocks are covered.

Section 2.3 presents more concepts in STA such as false paths, zero cycle, and multicycle paths as well as timing specifications. We end this chapter with a discussion of phase-locked loops.

2.2 Clock Definitions

The clock input is a periodic signal that controls all the timing characteristics in synchronous devices. A clock signal can be either positive or negative. The clock shown in Figure 2.1a is a positive clock pulse. Figure 2.1b represents a negative clock pulse.

In a positive clock pulse, the leading edge of the pulse makes a momentary transition from 0 to 1, whereas in a negative clock pulse,

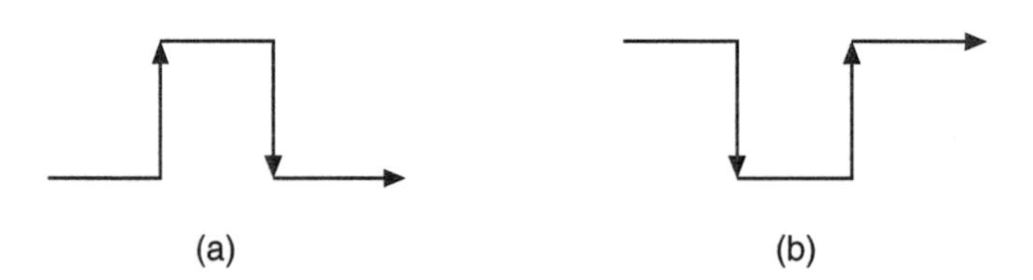

Fig. 2.1 Single Clock Pulse: (a) Positive Clock Pulse (b) Negative Clock Pulse

the momentary transition is from 1 to 0. In modeling clock behavior, it is common to assume that edges of the clock (rising and falling) are active when they reach 50% of the next value in transition. There is also uncertainty when clock-edge transitions are taking place. Uncertainties are defined as the difference between the earliest and the latest tolerated occurrence. This is shown in Figure 2.2.

Figure 2.3 shows the timing relation for a rising-edge triggered flip-flop (e.g., a positive clock pulse). Data propagates to output of the flip-flop at the rising edge of the clock after delay is introduced by flip-flop device. This delay is commonly called Clock to Q. In a similar fashion, Figure 2.4 shows the timing relation for a falling-edge triggered flip-flop (e.g., a negative clock pulse).

Due to different timing characteristics of each component, several clock inputs usually exist in a circuit. In this case a clock network, which is a distribution of a reference clock, or main clock, is available for the circuit. Figure 2.5 represents a clock network.

To distribute a clock signal, the following points must be taken into consideration:

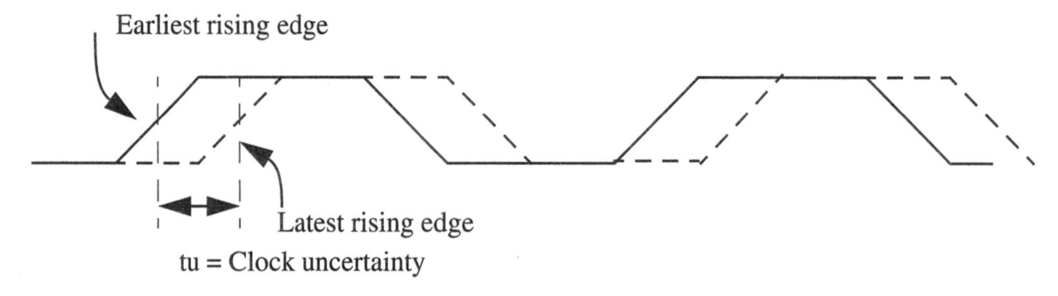

Fig. 2.2 Rising/Falling Edges and Uncertainties in a Clock

- Allow only one path to exist from the reference clock to storage elements—buffers or inverters.
- Discourage the use of clock dividers.
- Disallow any logical combination of the clock signals.
- Gate the clock signals only when done in a hazard-free manner.

Figure 2.6 is an example of a clock distribution structure.

2.2.1 Gated Clocks

A clock signal can be gated to produce a delayed clock signal. It is important to point out that clock gating can be done only on single frequency systems and multiple clocks cannot be combined to create a different frequency. Figure 2.7 represents a gated clock. It is an

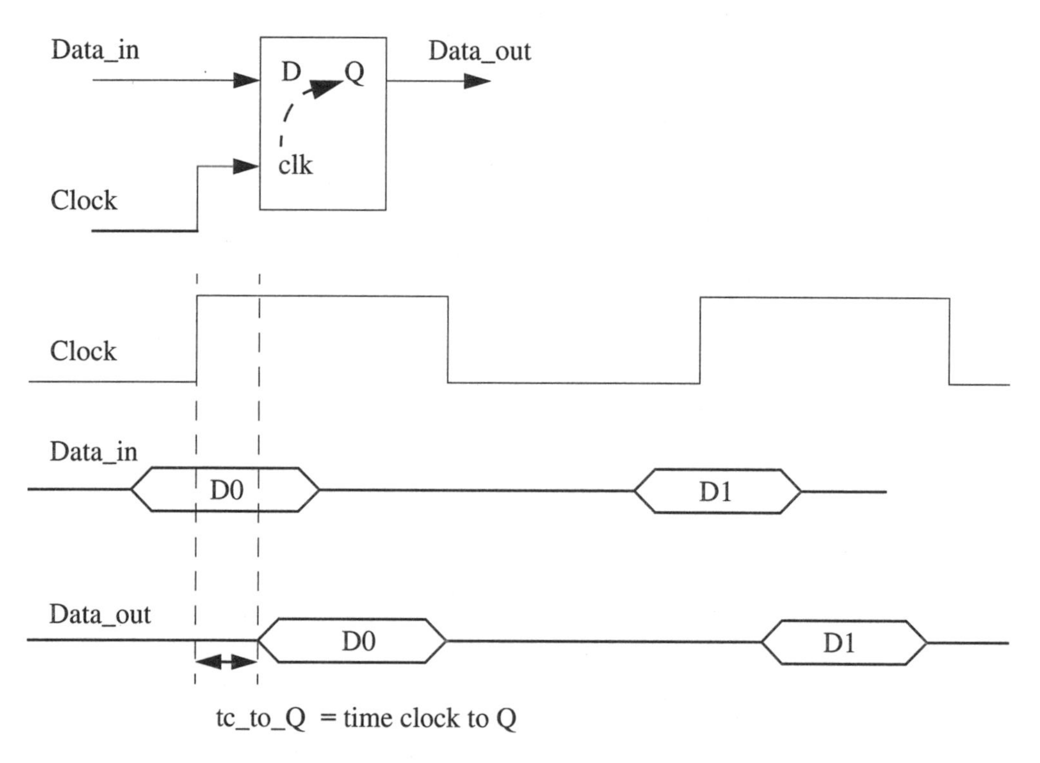

Fig. 2.3 Positive Clock Pulse in Relation with Rising-Edge Triggered Flip-Flop

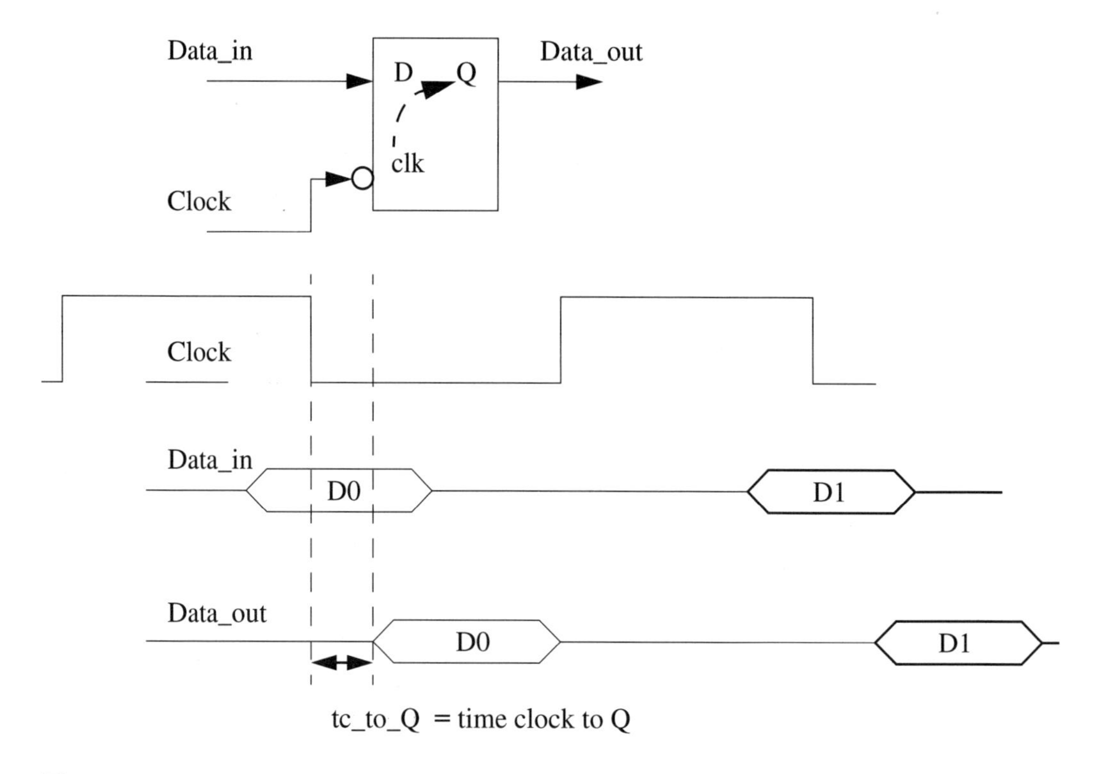

Fig. 2.4 Negative Clock Pulse in Relation with Falling-Edge Triggered Flip-Flop

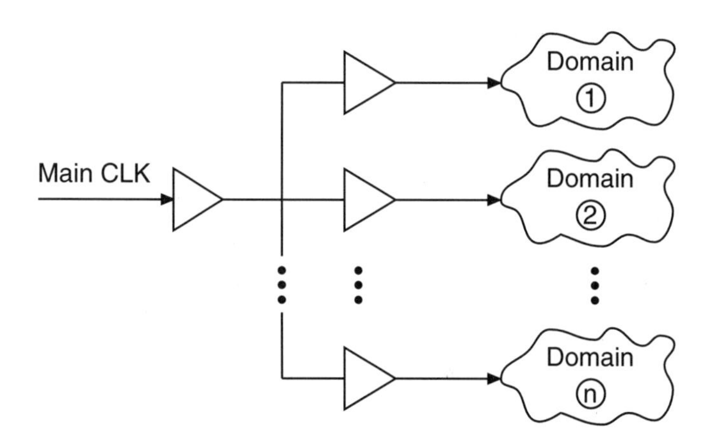

Fig. 2.5 A Clock Network

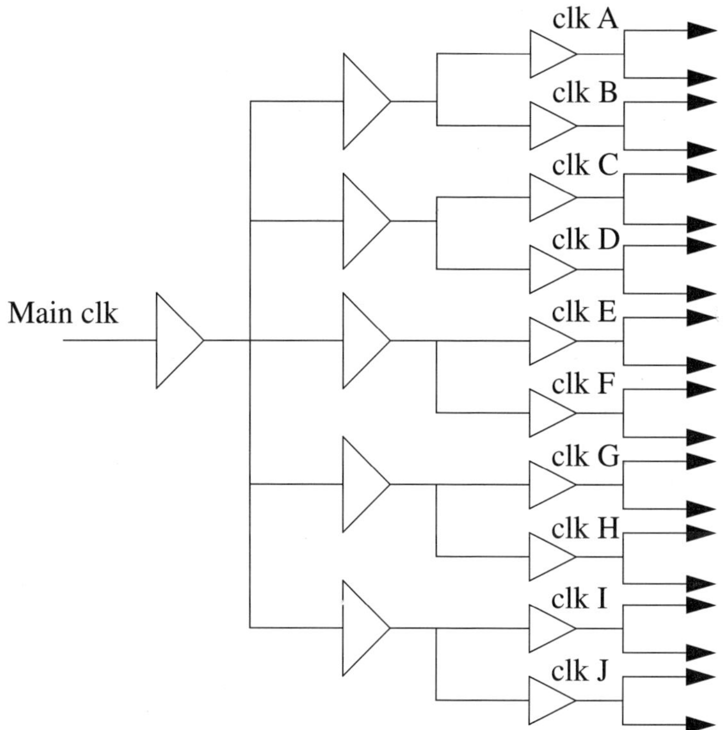

Fig. 2.6 Clock Distribution Structure

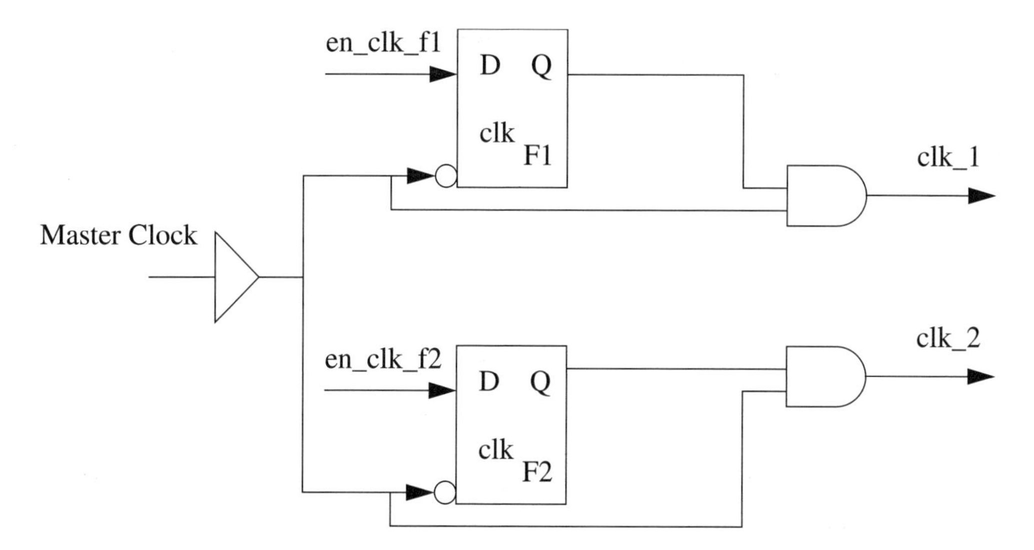

Fig. 2.7 Gated Clock

example of properly designed gated clocks (clk_1 and clk_2) from the master clock. Each of the clocks is enabled by a separate control signal (en_clk_f1 and en_clk_f2) sampled on the negative edge of the master clock. Both gated clocks are generated glitch free.

2.2.2 Clock Skews and Multiple Clock Groups

Clock skew is defined as the time difference between the clock-path reference and the data-path reference. The clock-path reference is the delay from the main clock to the clock pin and the data-path reference is the delay from the main clock to the data pin of the same device. There are four types of clock skew:

1. Polarity skew
2. Phase skew
3. Frequency skew
4. Cycle skew

The polarity skew happens when the active edges of the source and destination clocks are in opposite directions as shown in Figure 2.8. The figure shows one possible implementation of clock distribution in a digital pipelined synchronous design. Flip-flops of the pipelined stages are triggered with both positive and negative clock edges. The advantage of this clocking scheme is in efficient use of the clock. For the same clock frequency, throughput of data is doubled because data is moving on both clock edges.

Phase skew is the time difference between the active edges of the delayed source and destination clocks in the same cycle, as shown in Figure 2.9. Phase skew in clock distribution is very similar to the effect of clock skew introduced by physical design of the clock tree. Controlling delays in a clock tree in such fashion are difficult to design and analyze using static timing analysis and should be avoided.

Frequency skew happens when the source and destination clocks have different frequencies as shown in Figure 2.10. This is an example of interfacing between different clock domains. It is a good

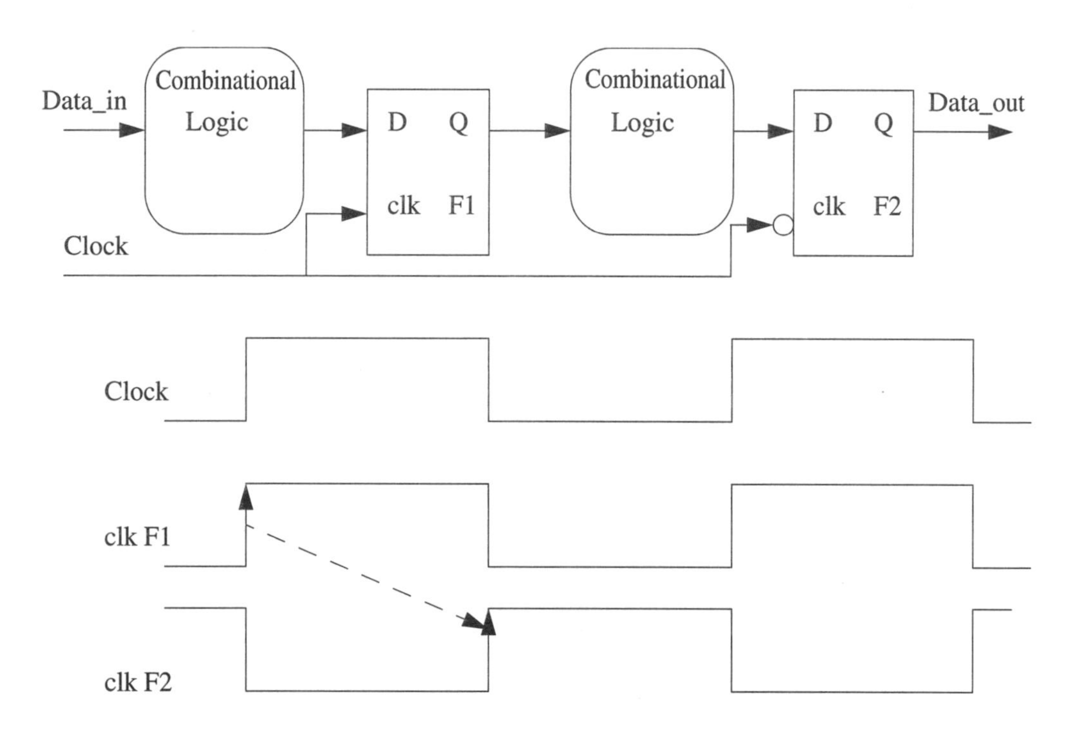

Fig. 2.8 Polarity Skew

practice to work with a single clock domain for digital synchronous designs. However, in some cases, it is inevitable to work with multiple clock domains. Such cases should be localized in a separate level of hierarchy and carefully handled in static timing analysis, possibly using multicycle paths.

Cycle skew corresponds to the number of cycles that the source clock has to shift to align with the active edge of the arriving destination clock. Figure 2.11 represents the cycle skew. This is an example of a two-clock cycle skewed design. Combinational logic 2 has a critical path requiring two clock cycles to propagate through. The critical path through combinational logic 2 is a multicycle path for static timing analysis. This design practice is common for cases where adding new pipeline stages and increased latency is not acceptable.

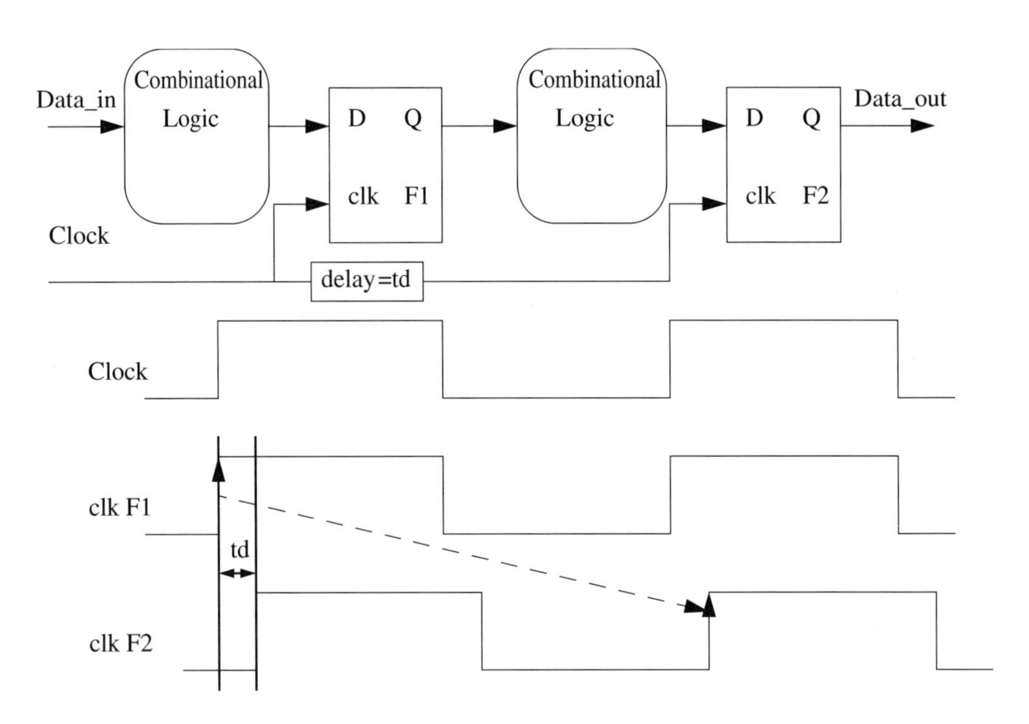

Fig. 2.9 Phase Skew

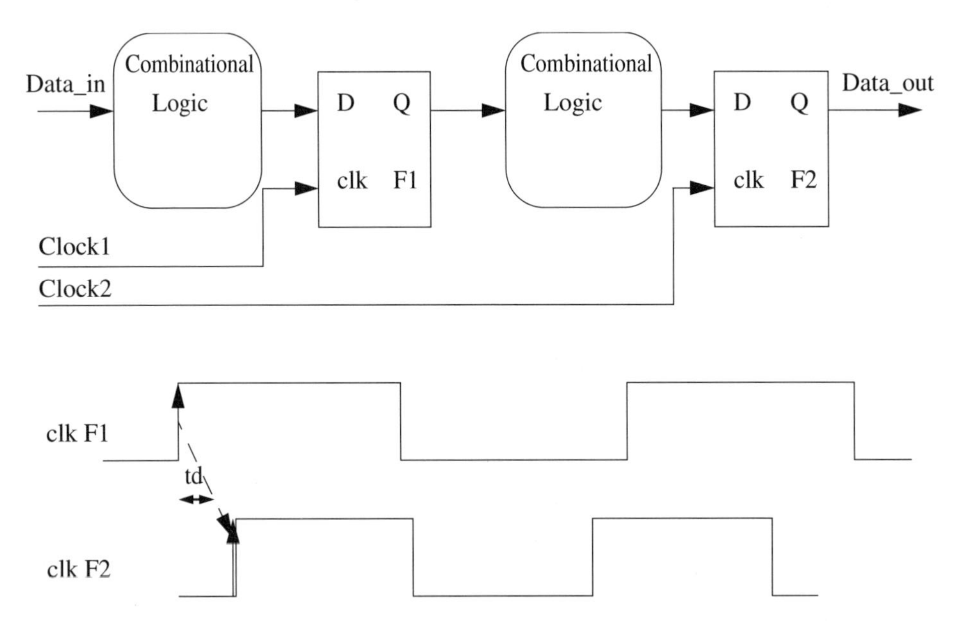

Fig. 2.10 Frequency Skew

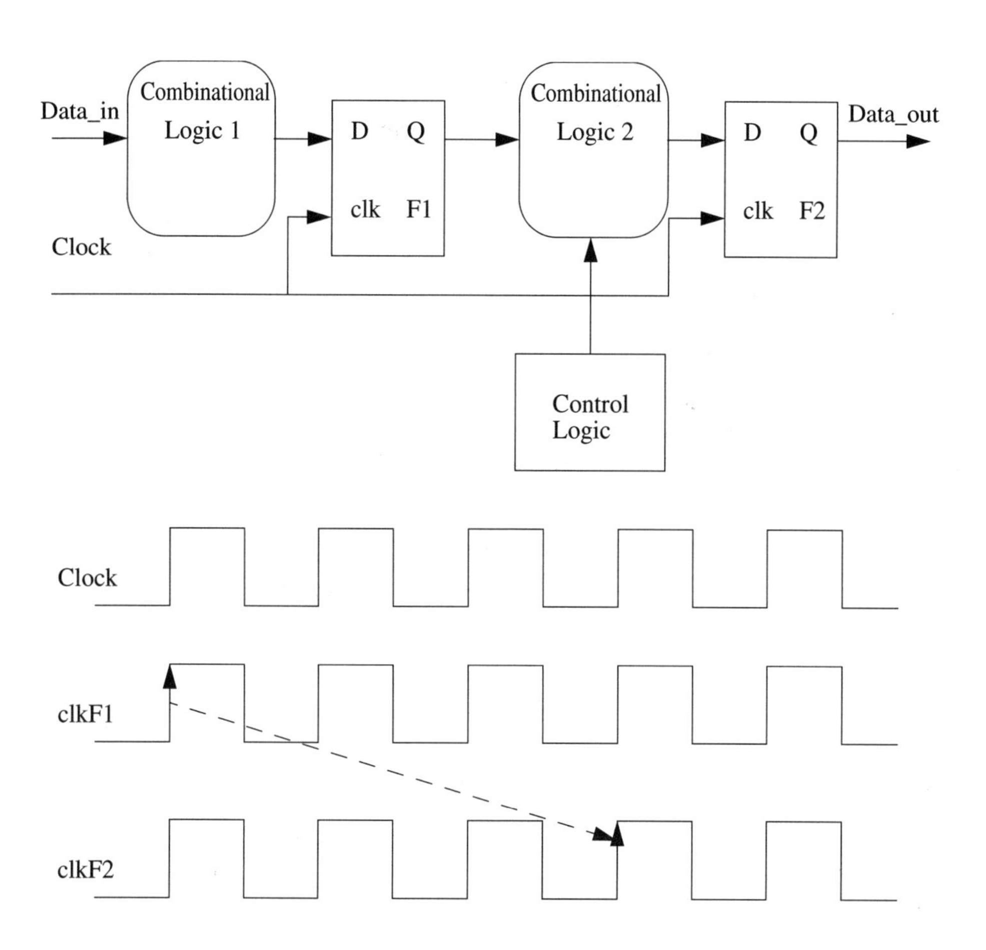

Fig. 2.11 Cycle Skew

2.2.3 Multifrequency Clocks

Most of the complex designs require multiple clock frequencies. The multifrequency clocks can all be generated from a reference, or main, clock. In that case the design will not be completely asynchronous which makes the timing analysis not difficult. Figure 2.12 shows a multifrequency clock network.

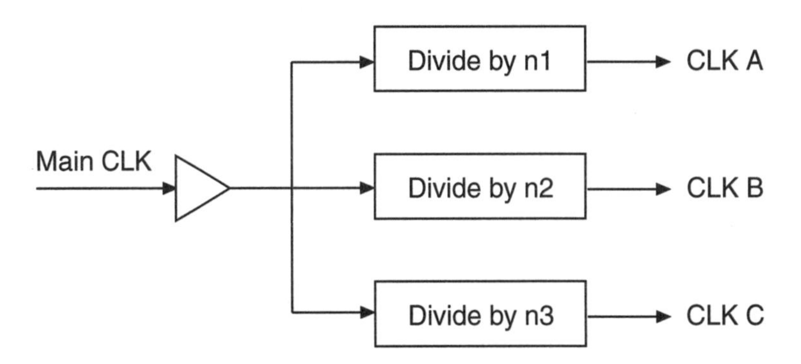

Fig. 2.12 Multifrequency Clock

2.2.4 Multiphase Clocks

Multiphase or delayed clocks are required in most of the designs. Figure 2.13 shows a multiphase clock network and its corresponding waveforms. This is a three-phase clock network. Latch-based circuits are commonly designed using a multiphase clock network.

As we mentioned in chapter 1, static timing analysis is a process to verify the timing constraints of a synchronous sequential design without checking its functionality. The constraints that are checked are setup and hold-time violations, recovery and removal violations, delay calculations, and minimum-pulse width and maximum-phase skew violations.

Setup time is the minimum time that the synchronous data must arrive before the active clock edge in a sequential element. Hold time is the minimum time that the synchronous data is stable after the active clock edge.

Recovery and removal are analogous to setup and hold times but for asynchronous control signals such as preset or clear. So recovery time is defined as the minimum time that an asynchronous control signal must be stable prior to the active edge of the clock. Removal time is the minimum time that an asynchronous control signal is required to be stable after the active edge of the clock.

The minimum-pulse width of a clock is the minimum duration that the clock has to be asserted to ensure the data is valid. The

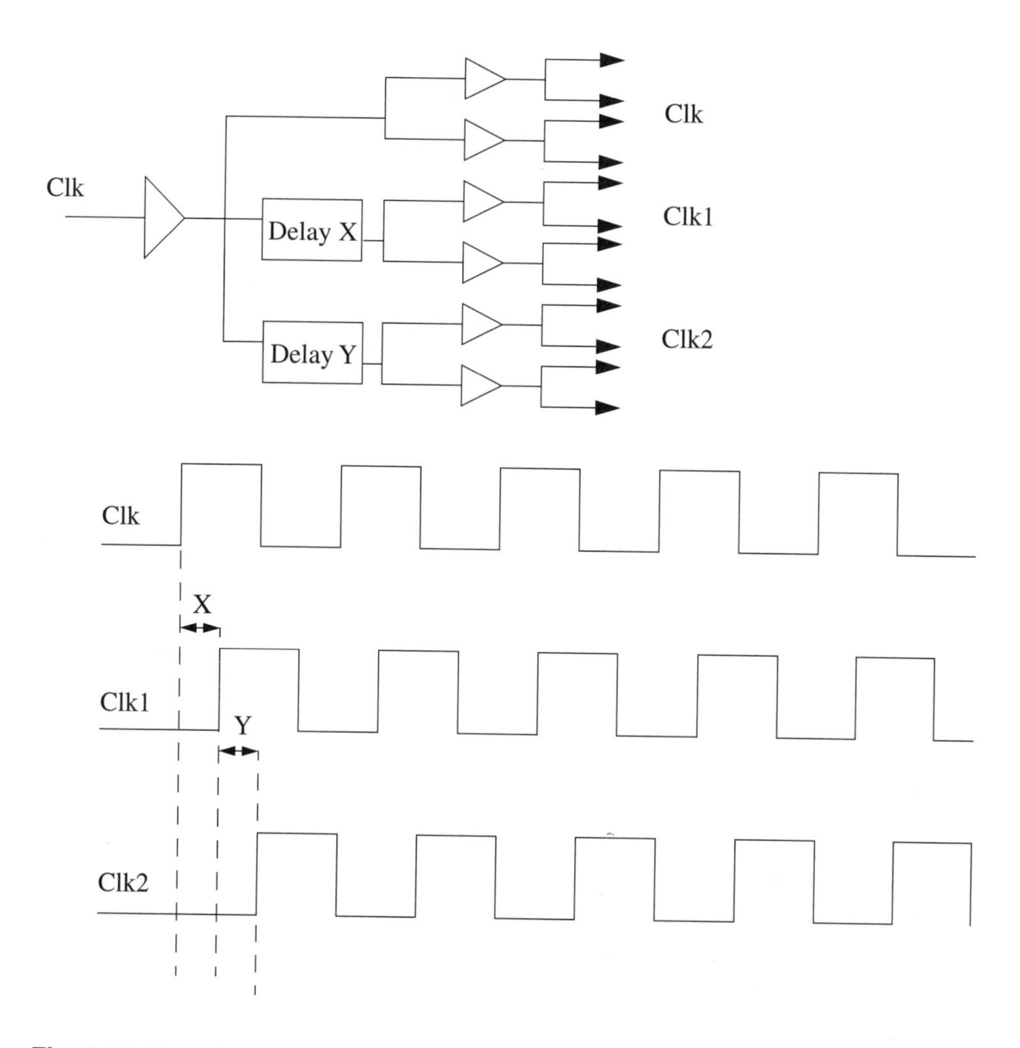

Fig. 2.13 Multiphase Clock

maximum-phase skew is the maximum time that two control signals of opposite phase overlap with each other.

Figure 2.14 shows the basics of setup and hold times. In this figure we have a generic pipeline stage of a synchronous design. Input data (Data_in) is processed in a combinational logic and the result is saved in a D flip-flop. Combinational logic has maximum delay path t_m. Data on input D must be stable for a minimum

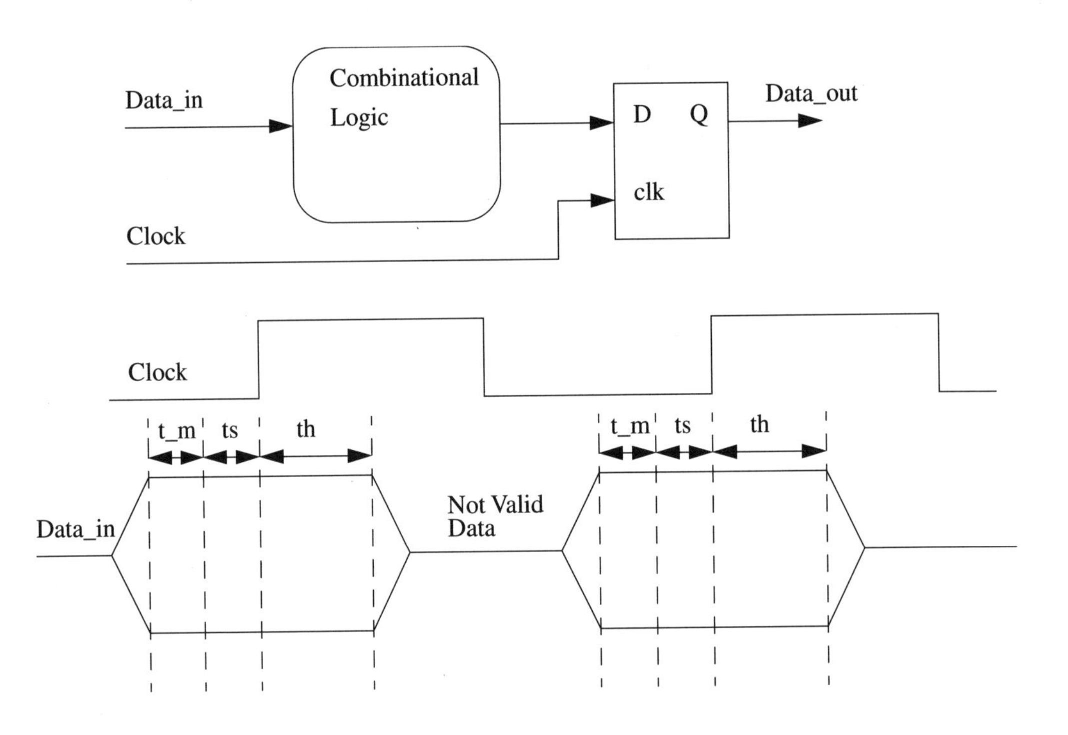

Fig. 2.14 Setup and Hold Times in Generic Pipeline Stage of Synchronous Design

length of time (ts is setup time) before the rising edge of clock. Data on input D must stay the same after the rising edge of clock for a minimum length of time (th is hold time).

The setup-time and hold-time constraints have separate rising and falling components; therefore, two paths must be considered in performing the static timing analysis.

2.3 More on STA

In chapter 1 we covered some basic concepts of STA. In this section we cover more topics in STA.

2.3.1 False Paths

As the analyzer determines delay, it considers only the paths that actually affect the output. If a path is never activated, or sensitized, it cannot possibly contribute to delay. Once the active paths are identified, the slowest of them fixes the block's propagation delay. The circuit in Figure 2.15 is the combination of two modules. The circuit on the right of the dashed line is the same as Figure 1.3. The new module modifies the function of the old circuit in a way that changes its timing. An analysis of the combined modules shows that the delay on line 3, of Table 4.4, has been eliminated. The longest delay path, a-out through c1-c2-c4-c5-c6, is now implemented through gates k1-k2-c6. The delays of lines 4 and 5 are also gone since the circuit no longer enters states 10000 or 10100.

Since the slowest path is no longer used, it no longer contributes to the delay and it is known as a false path. Many STA tools try all possible combinations on a module's input signals and look for paths that do not toggle. Paths that do not change are automatically marked as false paths and are not used in the delay analysis. The false path in Figure 2.15 may or may not be automatically detected by the software because, although it is bypassed, it still toggles. Some tools can also detect paths that are logically never used which describes path a-out in this case. The path c1-c2-c4-c5-c6 must be declared a false path or the module's timing will not be accurate.

The set_false_path command identifies false paths. The following statement means that any path that travels through n1 and n2 and n4 and n5 is a false path.

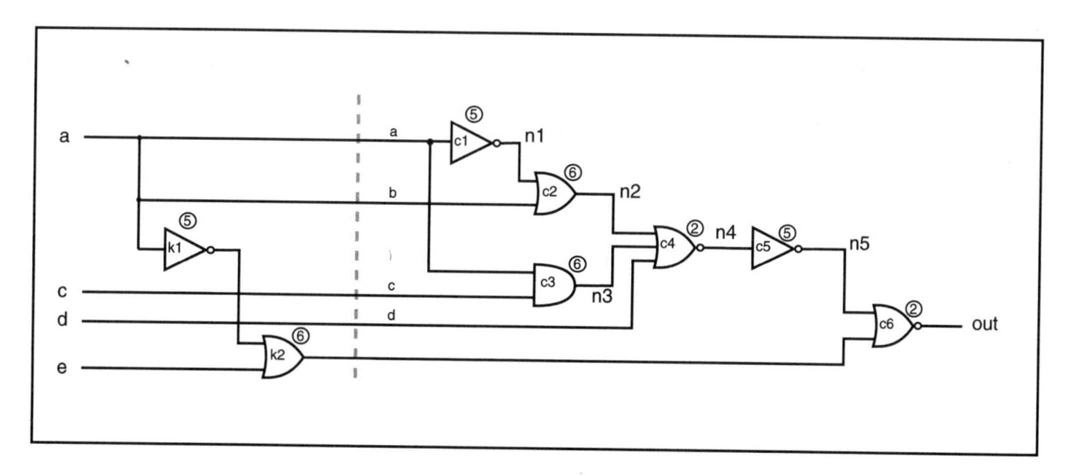

Fig. 2.15 Slight Modification to Circuit in Figure 1.3 Makes the a-out Path a False Path

```
set_false_path -through {n1} -through {n2}
                             -through {n4} -through {n5}
```

Very slow signals that are used only under special conditions and not during normal operation, like reset, test-mode enable, or constant values, can also be manually specified as false paths. Any path that does not change or does not affect the circuit's operation should also be labeled as a false path.

2.3.2 Multicycle Path Analysis

Multicycle paths in a design are the paths that intentionally require more than one clock cycle to become stable. Therefore, they require special multicycle setup and hold-time calculations. Multicycle paths are found in the designs that use both the positive and negative edges of a clock, and designs that use multiple tristate buses. These special paths must be identified, so the analyzer does not incorrectly interpret its operation. The timing in a circuit is defined by the clock period. All delays are measured against the ruler of the active clock edge. A circuit designed to take more than one cycle to do its work will always be marked as not functioning correctly. Valid paths, like those shown in Figure 2.16, that require more than one cycle are identified with the multicycle path command. Declaring a multicycle path tells the analyzer to adjust its measurements so it does not incorrectly report setup time violations. The multicycle statement shown in Example 2.1 declares a path from input data to output multval that requires three cycles to complete.

Example 2.1

```
Set_multicycle_path 3 -from data -to multval
```

Zero cycle paths occur when the data is latched into a state device and in the later clock cycles is transferred into another state device. This creates a race condition in the design. Figure 2.17 shows a zero cycle path. In this figure the data from the flip-flop is transferred to the RAM on the same clock cycle. Sometimes design-

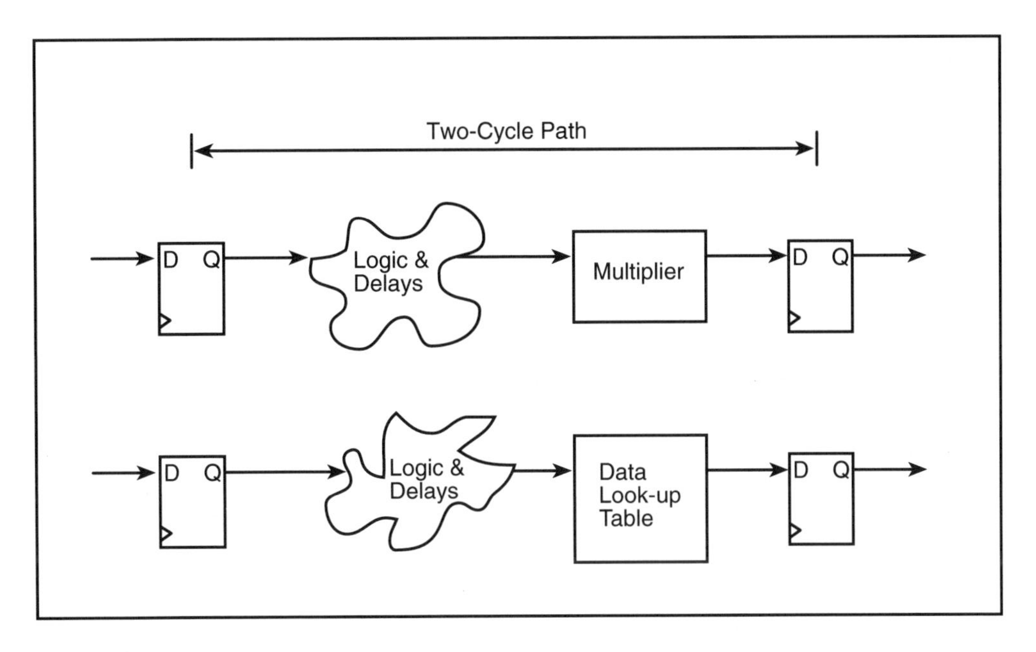

Fig. 2.16 Valid Paths That Take More Than One Cycle to Complete Must Be Identified as Multicycle Paths

ers intentionally create zero cycle paths. However, if these are not intentional, they should be flagged as hold-time violations.

2.3.3 Timing Specifications

The purpose of STA is to determine if the circuit can meet specified timing requirements. The designer sets some requirements while others are imposed by process limitations. Timing requirements decided upon by the designer are described below.

2.3.3.1 Clock Period, Skew, and Duty Cycle Using the clock specification, the timing analyzer can determine the amount of time allowed for propagation between registers. The impact of the setup time on allowed propagation delay was previously discussed; however, there is another factor that affects the propagation time: skew. The waveform shown in Figure 1.3 which is repeated here as Figure

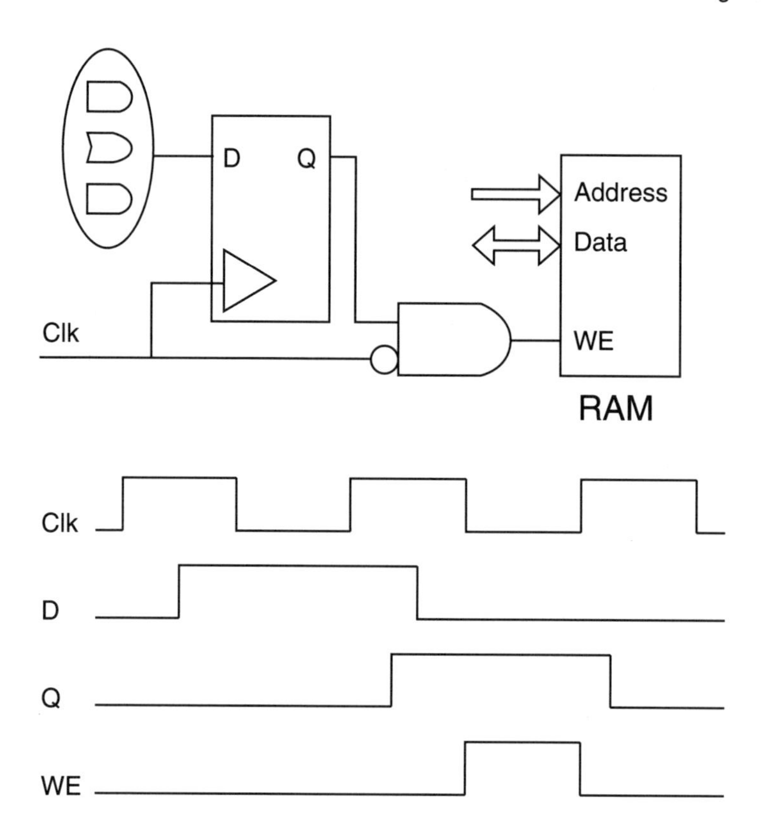

Fig. 2.17 Zero Cycle Path

2.18 shows how clock-skew and flip-flop setup time are subtracted from the clock period to arrive at the time left for propagation. Any path that takes longer than

$$Tprop = Tperiod - (Tskew + Tsetup) \quad \textbf{(Eq. 2.1)}$$

to propagate, violates the path's required timing. The clock and skew are specified in separate commands. The skew can be specified and the amount it affects setup and hold times as shown in Example 2.2.

Example 2.2

```
create_clock -period period -waveform {rise_edge, fall_edge} name
create_clock -period 15 clock -waveform {0, 7.5} clock

set_clock_uncertainty -[setup][hold] time signal
set_clock_uncertainty -setup 0.21 clock
set_clock_uncertainty -hold 0.12 clock
```

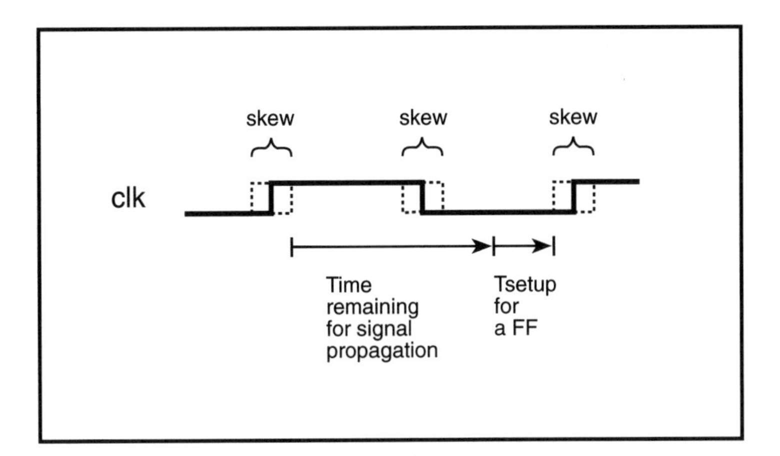

Fig. 2.18 Time Available for Propagation Between Two Flip-Flops Depends on Clock Skew and Flip-Flop Setup Time

Once the layout is completely done, the skew in the clock tree can be measured by the static timing analyzer instead of specified by the designer. The analyzer is instructed to perform the clock tree analysis with the set_propagated_clock command.

A clock's duty cycle does not have to be 50–50. It can be defined as any percentage high or low. The analyzer can check to make sure the clock's pulse width never goes below a set minimum. The set_min_pulse_width_check command ensures that the active high and active low parts of the signal do not slip below set minimum widths. In Example 2.3, the clock signal must be low for at least 3.3ns and high for 4.2ns; otherwise an error is flagged.

Example 2.3

```
set_min_pulse_width_check -low 3.3 -high 4.2 clock
```

2.3.3.2 Process Characteristics The physical properties of a fabrication process vary from one lot to the next. Process variations directly affect the propagation delays of gates and interconnect. The designer can choose between worst-case, best-case, and typical-process parameters and conditions for the timing analysis. The set_operating_conditions command specifies the parameters to use. The command in Example 2.4 specifies the best-case commercial process for minimum and worst-case commercial process for the maximum propagation delays.

Example 2.4

```
set_operating_conditions -analysis_type bc_wc
                                     -min BCCOM -max WCCOM
```

2.3.3.3 Module Input and Output Delay An additional delay can be added to the input and output of a module to account for interconnect delay that is not part of the net list or to represent delays through I/O pads. Example 2.5 shows statements to set both input and output delays for the circuit shown in Figure 2.15. An additional 3.3ns is added to all the inputs a through e. A delay of 1.6ns is added to the output.

Example 2.5

```
set_input_delay delay_value pin_name
set_input_delay 3.3 {a b c d e}

set_output_delay delay_value pin_name
set_output_delay 1.6 out
```

2.3.3.4 Back Annotation The standard delay format (SDF) is used to pass delay information between synthesis, STA, and layout tools. The designer must specify which files to use in the analysis with the commands in Example 2.6. The word *filename* means the name of the file that contains the parasitic or SDF information.

Example 2.6

```
read_parasitics filename
read_sdf filename
```

2.3.3.5 Delay Model Analyzers can also accept parasitic delay information files that detail the capacitance and resistance values of interconnect. Two models can be used for timing analysis: reduced standard parasitic format (RSPF) and detailed standard parasitic format (DSPF). Both models are shown in Figure 2.19. The reduced model splits an entire interconnect line into two capacitors and a resistor. The detail model represents each line segment between gates as a resistor and capacitor. Modeling interconnect manually will be discussed in section 3.3. The detailed model is the most accurate and if the information can be extracted for the detailed model, manual extraction and modeling will not be necessary. The type of parasitic delay model used depends on the type of data given the analyzer through the read_parasitics command as given in Example 2.6.

2.3.4 Timing Checks

Once the circuit's expected performance is defined, the STA tool can check its performance. The more accurately and completely the tim-

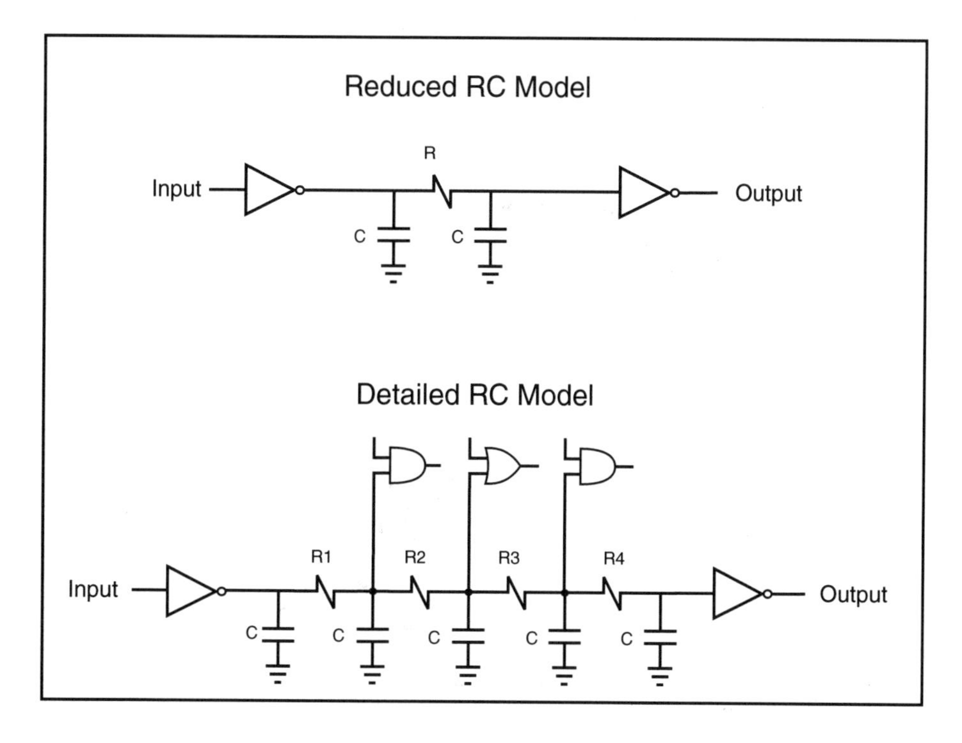

Fig. 2.19 CAD Tools Offer Two Delay Models

ing requirements are specified, the fewer false errors will be generated. The most common timing checks are listed below.

2.3.4.1 Maximum Delay A maximum delay can be set for any signal including a clock. If a signal must arrive at the input of a gate in a certain amount of time, the analyzer will monitor the signal and report any violations. The delay is set with reference to the clock. The command format and its use are given in Example 2.7 where a delay of 7.8ns is set as the limit for any signal to reach the output data_out.

Example 2.7

```
set_max_delay delay -to pin
set_max_delay 7.8 -to data_out
```

2.3.4.2 Minimum Delay Minimum delay is the inverse of the maximum delay. It sets a lower boundary on how fast a signal may propagate to a certain point. Once the minimum delay is specified, the analyzer will report any violations. The minimum delay set in Example 2.8 is 4.3ns to the output r_out.

Example 2.8

```
set_min_delay delay -to pin
set_min_delay 4.3 -to r_out
```

2.3.4.3 Setup and Hold Times If the setup and hold times of a flip-flop are not met, the circuit will fail. With gate-level simulations, one cannot be absolutely sure that all possible paths to flip-flops are tested. STA removes any uncertainty because it can guarantee that all paths to all flip-flops are measured and tested for setup and hold time violations. Setup and hold times are determined by the characteristics of the gates in the technology library. The analyzer automatically finds the necessary times when it reads in the library and verifies that they are not violated.

2.3.4.4 Recovery and Removal Times Recovery and removal times are similar to setup and hold times except they verify the timing of asynchronous control signals, like clear and set, with relation to the clock. The timing requirement is determined by the flip-flops in the technology library.

2.3.4.5 Minimum Clock Width The duty cycle of the clock is verified by checking the minimum clock widths. Specifying a minimum width also allows the analyzer to check for spikes or glitches in the clock. If the clock is not gated, but comes directly from the clock tree to each register or latch, there will be no glitch problems; however, if any logic intervenes between the clock and the registers, a devastating glitch can occur. It is a good design practice to not gate the clocks, but if your design demands it, use the set_min_pulse_width command to verify that there are no glitches. The command to check for minimum pulse widths is shown in Example 2.9 where the minimum allowed width for any pulse on clock is 15ns.

Example 2.9

```
Set_min_pulse_width minimum_time signal_name
Set_min_pulse_width 15 clock
```

2.3.4.6 Minimum and Maximum Clock Skew Synchronous design defines everything with relation to the clock. If the timing of the clock is not consistent throughout the entire circuit, some timing requirements may not be met, especially setup and hold times. If the data timing remains constant, but the clock moves ahead or back, setup or hold violations may occur. Skew is inherent and unavoidable in clock trees, but the analyzer can ensure that it stays within manageable boundaries.

Most standard cell projects use a custom clock tree designed by the vendor. The tree has guaranteed minimum and maximum skew between branches. The static timing analyzer needs to know the skew numbers. Before layout, the skew is simply an estimate. After layout, the timing tool can extract the skew from the SDF file, so it does not need to be explicitly provided.

2.3.4.7 Point-to-Point Propagation Static timing analyzers can report the propagation delay between any two nodes. This capability makes it possible to analyze asynchronous circuits when necessary or the delay through an input pad all the way through to an output pad.

2.4 Timing Analysis of Phase-Locked Loops

Most ASICs currently developed include one or more phase-locked loop (PLL) circuits. PLLs are used for a number of reasons including reduction of on-chip clock latency, synchronization of clocks between different ASICs, frequency synthesis, and clock frequency multiplication.

2.4.1 PLL Basics

A simplified block diagram of a PLL is shown in Figure 2.20. The reference clock, REFclk, is the input clock into the ASIC. The PLL tracks the reference clock and adjusts the phase of its output, PLLout, such that REFclk and the feedback clock, FBclk, are in phase.

The phase detector compares the phase difference between the rising edge of REFclk and FBclk. When the two are not aligned, the

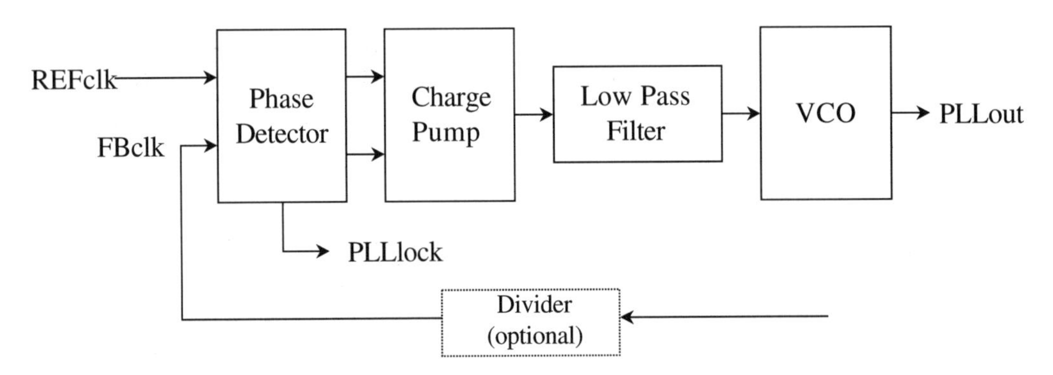

Fig. 2.20 Phase-Locked Loop Block Diagram

phase detector output changes to increase or decrease the voltage level on the output on the charge pump.

The charge pump and low-pass filter convert the digital output of the phase detector into an analog voltage. The low-pass filter is used to control rate of change of input voltage into the voltage controlled oscillator (VCO).

The VCO generates the clock output. The clock frequency changes as a function of the output of the charge pump.

If the PLL is being used as a frequency multiplier, the FBclk frequency is divided before being fed into the phase detector.

2.4.2 PLL Ideal Behavior

A PLL will adjust the phase of its output such that its reference input REFclk and its feedback clock are perfectly aligned, or in phase.

Consider the ASIC PLL circuit shown in Figure 2.21. In this ideal circuit, the PLL will perfectly align the arrival time of the feedback clock, tFB, with the arrival time of the reference clock, tREF (tFB = tREF). The output of the PLL is distributed throughout the ASIC through the use of a clock distribution network that is perfectly balanced such that the delay from the PLL output to every ASIC register is equal (dlya = dlyb).

The arrival time of the PLL output can be expressed as:

$$tPLLout = tFB - dlyb \quad \textbf{(Eq. 2.2)}$$

Since in this ideal case tREF = tFB and dlya = dlyb,

$$tPLLout = tREF - dlya \quad \textbf{(Eq. 2.3)}$$

or

$$tPLLout + dlya = tREF \quad \textbf{(Eq. 2.4)}$$

and the arrival time of the clock at a register, rega, can be expressed as:

$$tA = tPLLout + dlya = tREF \quad \textbf{(Eq. 2.5)}$$

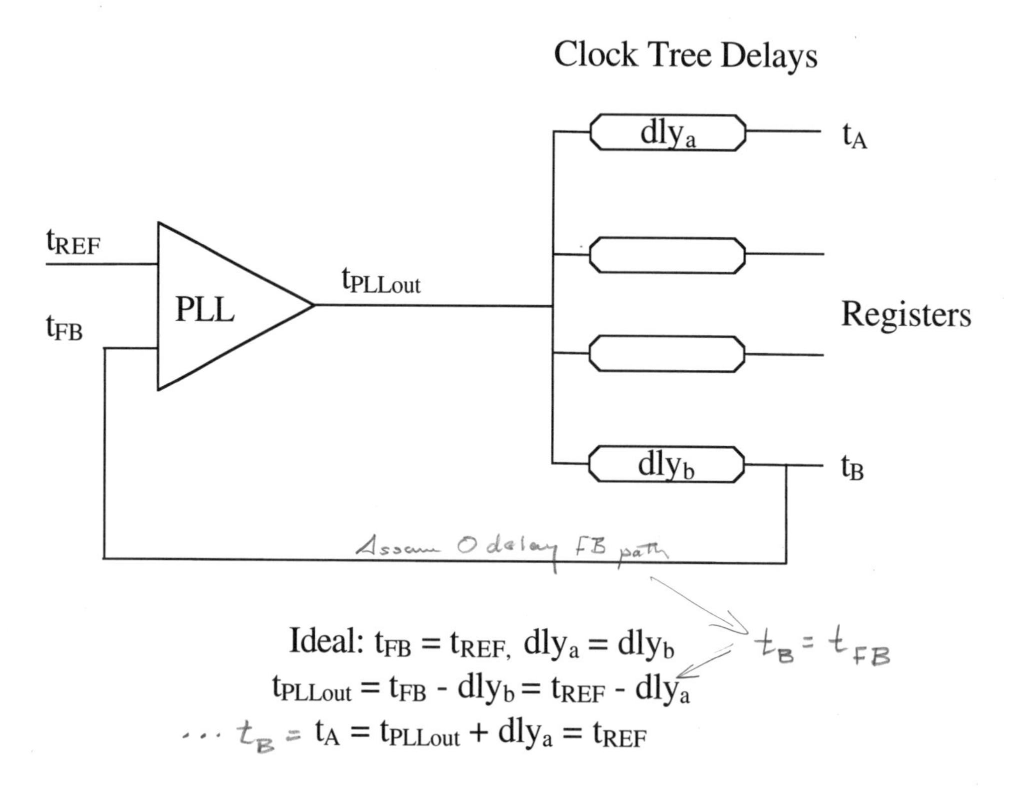

Fig. 2.21 Ideal PLL Behavior

So in this case, the arrival time of a clock at any register is perfectly aligned with tREF. Note that the delay of the clock distribution network is not a term of the arrival time, tA. This shows how a PLL can be used to minimize the effects of on-chip clock latency.

Consider also two ASICs fed by the same clock, but whose clock distribution networks have different latency times. Using PLLs in each of the ASICs will make the clocks in each of the ASICs appear to be in phase with each other.

$$ASIC1(tA) = tREF \quad \textbf{(Eq. 2.6)}$$

And

$$ASIC2(tA) = tREF \quad \textbf{(Eq. 2.7)}$$

hence,

$$\text{ASIC1(tA)} = \text{ASIC2(tA)} \quad \textbf{(Eq. 2.8)}$$

2.4.3 PLL Errors

Alas, neither PLLs nor ASICs exhibit ideal behavior. A PLL will not perfectly align tFB and tREF and the delays through an ASIC clock distribution network are never perfectly balanced. For purposes of timing analysis, the PLL errors to consider are listed below.

2.4.3.1 Static Phase Error SPE is the fixed offset (i.e., the error) between the rising edges of REFclk and FBclk caused by delay path differences in the phase detector, process/voltage/temperature differences with the ASIC, and transition time, or slew-rate, differences on the REFclk and FBclk inputs to the PLL.

2.4.3.2 Long-Term Jitter LTJ is a low frequency drift in the offset between the rising edges of REFclk and FBclk.

2.4.3.3 Short-Term Jitter STJ is a high frequency drift in the offset between the rising edges of REFclk and FBclk. This offset may vary from clock cycle to clock cycle. The STJ is a component of LTJ. For any given rising edge of PLLout, the next rising edge of PLLout will occur one clock period ± STJ later.

Jitter is usually caused by power and ground noise introduced into the VCO.

These PLL errors are expressed as ± delays. This means that because of nonideal behavior, tFB may occur before or after tREF. Typical values for PLL errors are:

Static Phase Error ± 200 ps max

Long-Term Jitter ± 250 ps max

Short-Term Jitter ± 100 ps max

Consider the ASIC PLL circuit previously analyzed (see Figure 2.22). In this case, however, nonideal behavior is introduced such

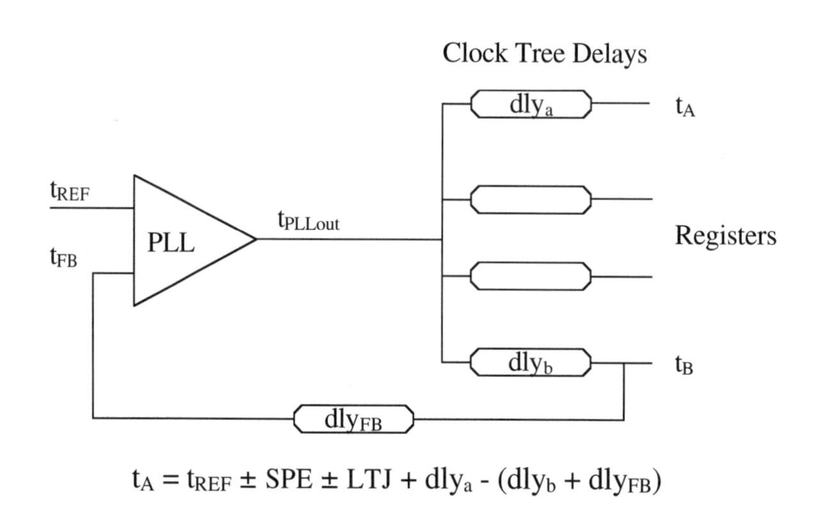

$$t_A = t_{REF} \pm SPE \pm LTJ + dly_a - (dly_b + dly_{FB})$$

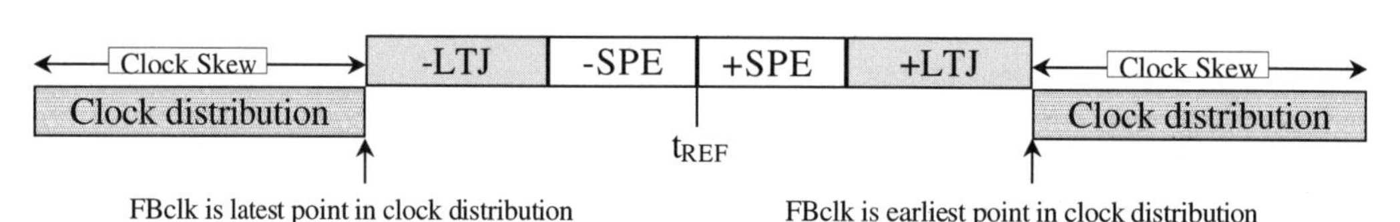

Fig. 2.22 Nonideal PLL Behavior

that the PLL exhibits SPE, LTJ, and STJ errors, and the ASIC clock distribution network is not perfectly balanced (i.e., the clock skew is nonzero). (Since STJ is a component of LTJ, STJ is not considered when analyzing the arrival time of a clock to any register, regA. The effects of STJ will be discussed later.)

The clock of regB is used as the feedback clock (FBclk) into the PLL. A delay exists between the arrival time of the clock at regB, tB, and the arrival of the clock at the FB pin of the PLL, tFB.

The difference between the arrival time of the clock at regA and the clock at regB represents the clock skew in the clock distribution network.

The arrival time of the clock at regA can now be analyzed as follows:

$$tFB = tREF \pm SPE \pm LTJ \qquad \textbf{(Eq. 2.9)}$$

tB = tFB – dlyFB **(Eq. 2.10)**

tA = tB – dlyB + dlyA **(Eq. 2.11)**

Substituting and rearranging,

tA = tREF ± SPE ± LTJ – dlyFB + (dlyA – dlyB) **(Eq. 2.12)**

The greatest errors occur if the arrival time of the FBclk, tB, is either the earliest or the latest point in the clock distribution. The arrival time, tB, should be chosen as the midpoint in the clock distribution network if the error between tREF and tA is to be minimized.

If intentional skewing of tA relative to tREF is desired, the feedback path delay, dlyFB, can be adjusted to achieve the desired effect.

Consider again two ASICs with PLLs driven by the same clock REFclk with clock skews of ASIC1(skew) and ASIC2(skew). Because of the errors introduced by the PLLs and the clock distribution networks on each ASIC, the arrival time of clocks with the ASIC may vary by as much as SPE+LTJ+ASIC1(skew)+ASIC2(skew). This can be a considerable amount, affecting the ability to clock signals from one ASIC to the next, if the clock skews within each ASIC are not carefully controlled and if the FBclk is not carefully chosen.

SPE and LTJ affect the arrival time of a clock to a register relative to REFclk. This means that setup and hold time of ASIC I/Os affected by SPE and LTJ. Short-term jitter (STJ) effectively reduces the clock period, thereby requiring the ASIC to run at a higher frequency than required by REFclk. This must be taken into account when analyzing the critical path within the ASIC. For example, if the clock period of REFclk is cycREF, the ASIC must be analyzed using a clock period of (cycREF – STJ). If a multicycle path is analyzed using a period of n*cycREF, the use of a PLL requires that the same path run at n*(cycREF – STJ) or (n*cycREF – n*STJ).

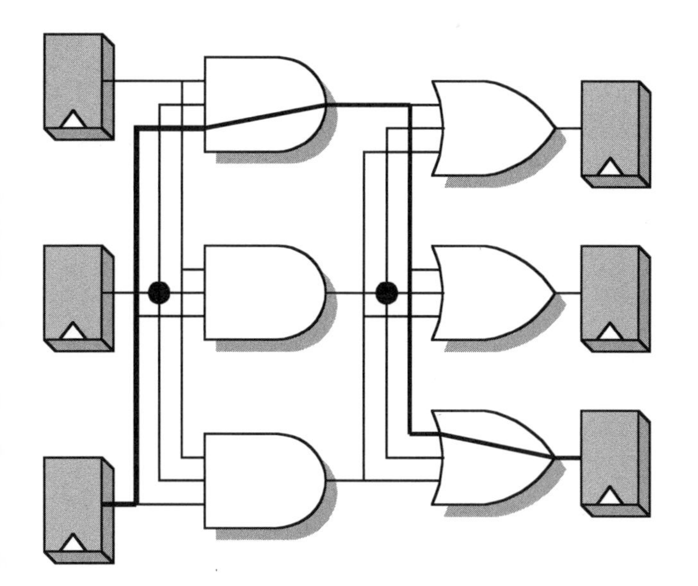

3 Timing in ASICs

3.1 INTRODUCTION

The number of ASICs designed increases every year. Advances in technology allow more transistors to be packed onto a single die which expands the applications where they can be used and accelerates development. Successful development of an ASIC depends on accurate modeling of its operation. Designing a circuit to be logically correct is simple. Producing an accurate timing model is critical to successful development. Current methodologies for generating accurate timing models for ASIC designs are described here.

Integrated circuits start as computer representations of a physical device. The designer's goal is to model the device characteristics with sufficient accuracy that actual silicon behaves as the model predicts, assuming the computer simulations exercise the model in the same way the device is expected to operate in the real world. Modeling a device's logical operation is relatively simple, and the translation from the model to the physical would be easy if it were not for the major difference introduced during fabrication: timing delays. The conversion of a logic statement to a model of its physical implementation is shown in Figure 3.1. The operation of the circuit

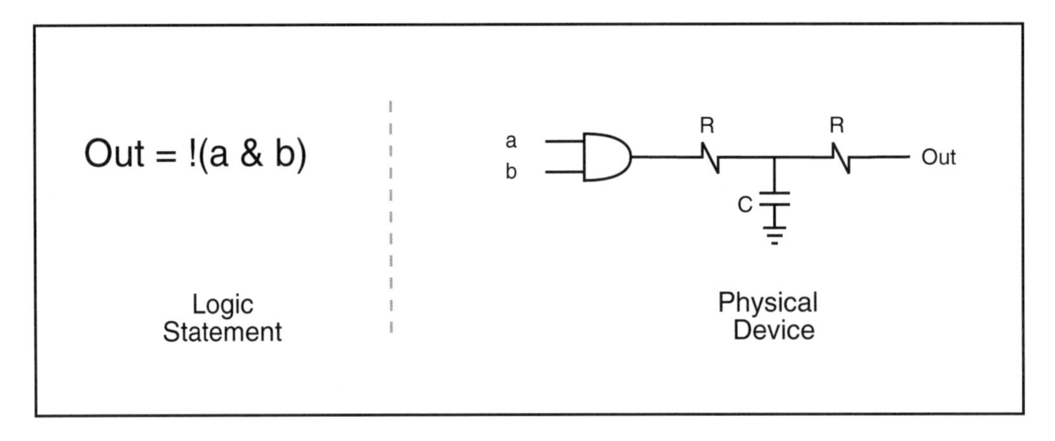

Fig. 3.1 Accurate Models Require Inclusion of Parasitic Capacitors

in Figure 3.1 is affected by the charging and discharging of the parasitic capacitor through resistors, both of which are inherent to silicon physical implementation. The stray capacitance and resistance can have such a great and deleterious effect that the physical operation is nothing like the simulated logical model.

A circuit's correct operation can be assured only if the timing of the simulated model is a close approximation of the final device. The accurate modeling of delay is of major importance. As process geometry shrinks and the number of transistors per die increases, the task of modeling the effects of parasitic capacitance and resistance makes it more challenging to correlate prelayout to postfabrication timing. Fortunately, CAD tools exist to accurately estimate delays before layout and extract the capacitance and resistance once layout is complete. Modeling estimated and extracted delays plays an important part in guaranteeing the timing and operation.

Any delay value used before the device is fabricated is merely an estimate. The four sources of delay are shown in Figure 3.2. Gate delay is determined by input slew rate and the inherent RC loading of the gate. Delay through a line depends on the RC load the gate drives. The fanout load simply increases the capacitance the driver must charge and discharge. Methodologies for predicting delay are well established. Gate delay is measured from fabricated test struc-

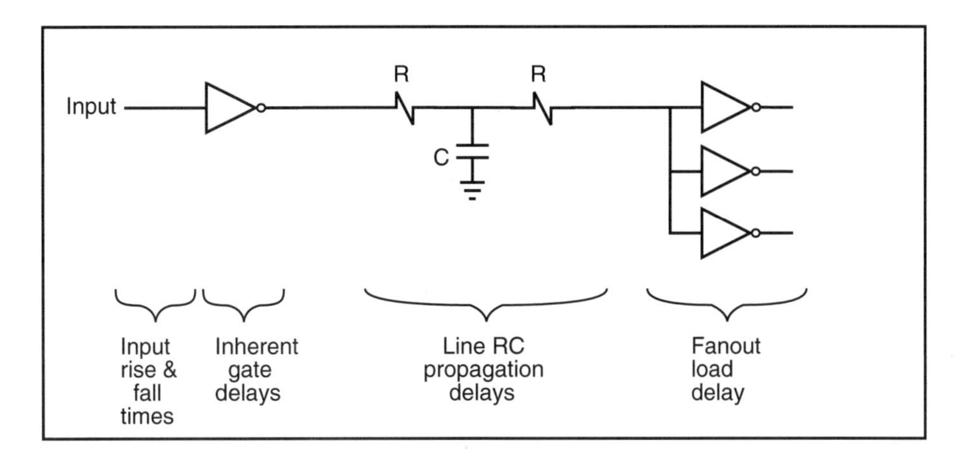

Fig. 3.2 Components of Circuit Delay: Input Slew Rate, Inherent Gate Delay, Line Propagation Delay, Fanout Load

tures tested at specific operating points. A transistor's speed, and therefore the inherent delay in a gate, is affected by its dimensions, the supply voltage, doping levels, input slew rate, operating temperature, and fanout load. The data measured from the test structure provides a device model that extrapolates to estimate delay under all operating and fabrication conditions.

The delay due to signal lines may be modeled in two stages: prelayout and postlayout. In either case, the physical characteristics of fabricated traces are known, having been measured from test structures. In the absence of layout, the unknown elements that affect timing are the trace's length, width, and surrounding signals. Figure 3.3 shows the parasitic capacitors seen by a metal trace. Parasitic capacitance is explored in detail in section 3.3. Before the layout is completed, any delay attributed to a signal line is an estimate based on probable length and width of the trace. Since the actual path is not known, the length is simply a guess based on the size of the overall circuit and the probability of placing the output of one gate close to the input terminals of the gates it drives. Another unknown aspect of the trace is the topology over which it passes. Once the layout is finished, the trace lines, and therefore their

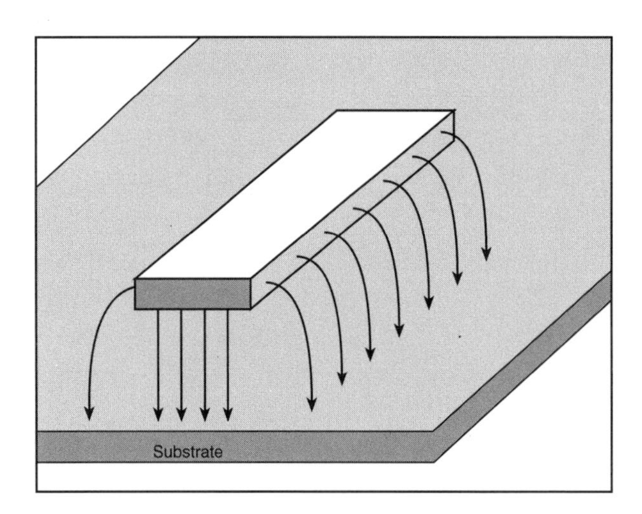

Fig. 3.3 Parasitic Capacitance of a Metal Line

delay, can be accurately modeled. The layout fixes their length and reveals what lies under the trace, whether it is substrate, transistors, or other layers.

Delay estimations are made in all stages of design: prelayout, synthesis, and postlayout. The most common methods used to estimate delays at all stages of the design cycle are explored.

3.2 PRELAYOUT TIMING

The design environment and methodology determine the accuracy and ease of modeling delays before the layout is finished. HDL languages, such as Verilog and VHDL, make it possible to add both gate and interconnect delays; however, except in situations where the layout is regular and known, such as in memories or decoders, the effects of delays due to interconnect are ignored until after synthesis or layout. The ease of including gate delays also depends on the type of model used. HDL modeling can also be done at two different levels: RTL and gate level.

3.2.1 RTL vs. Gate-Level Timing

RTL code models a logic function without regard to its implementation, whereas gate-level code specifies the exact gates required. Both the RTL and gate-level code for the logic function shown in Figure 3.4 are given in Example 3.1.

Example 3.1

RTL Code

```
out = ((a & b) | !a);
```

Gate-Level Code

```
and(a, b, s2);
not(a, s1);
or(s1, s2, out);
```

Modeling delay at the gate level is straightforward. The delay of each gate is found in the technology library. The appropriate delay can be assigned to every gate in the code and the propagation delay of signals estimated to provide a fairly accurate representation. However, manually implementing HDL code with delays for each gate is time consuming. At the prelayout stage, most design methodologies use synthesis to provide a gate model with delays while RTL code is used to model the circuit's behavior.

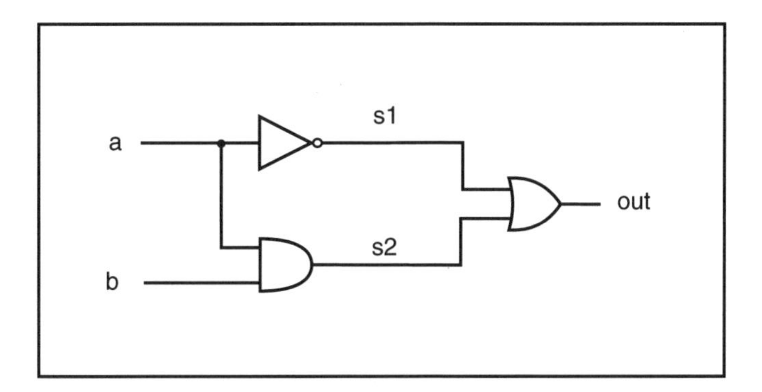

Fig. 3.4 Circuits Can Be Represented as RTL Code or Instantiated Gates

A clear method of accounting for delay is to determine the delay through each gate. The technology library already has delay information for every gate. Accurate modeling requires the assignment of the appropriate delay to each gate as described below. Estimating the delay of the RTL code is more difficult because of its level of abstraction. Until synthesis is complete, there is no straightforward way to correlate RTL code to actual gate delays.

The level of coding used affects the delays that can be modeled. Generally, RTL code is used to determine correct logical operation without regard for delays. A design at the gate level not only checks for correct operation, it also ensures that delays meet the required timing. Most designs start with an RTL code, then use synthesis to generate the gates needed to verify timing. Furthermore, few designs start at the gate level because the simulations, especially when timing is included, are very slow. Design at the RTL level offers a fast method to ensure that the logic is properly implemented. Synthesis then converts the design to gates that include delays from gates, estimated routing, and fanout.

3.2.2 Timing in RTL Code

Although it is impractical to assign delays to individual lines of RTL code, it is feasible to assign delays to entire modules. In RTL code, timing should be applied to any module or port that has a known response such as:

- Bus models
- Memories
- I/O ports
- Setup and hold times

A high-level system is shown in Figure 3.5. Each block is implemented as RTL. The RAM and the EPROM will not be synthesized. They are both modeled as an array of memory indexed by the address. The processor comes from a vendor's library. Its model reflects only the bus transactions that take place. The address-

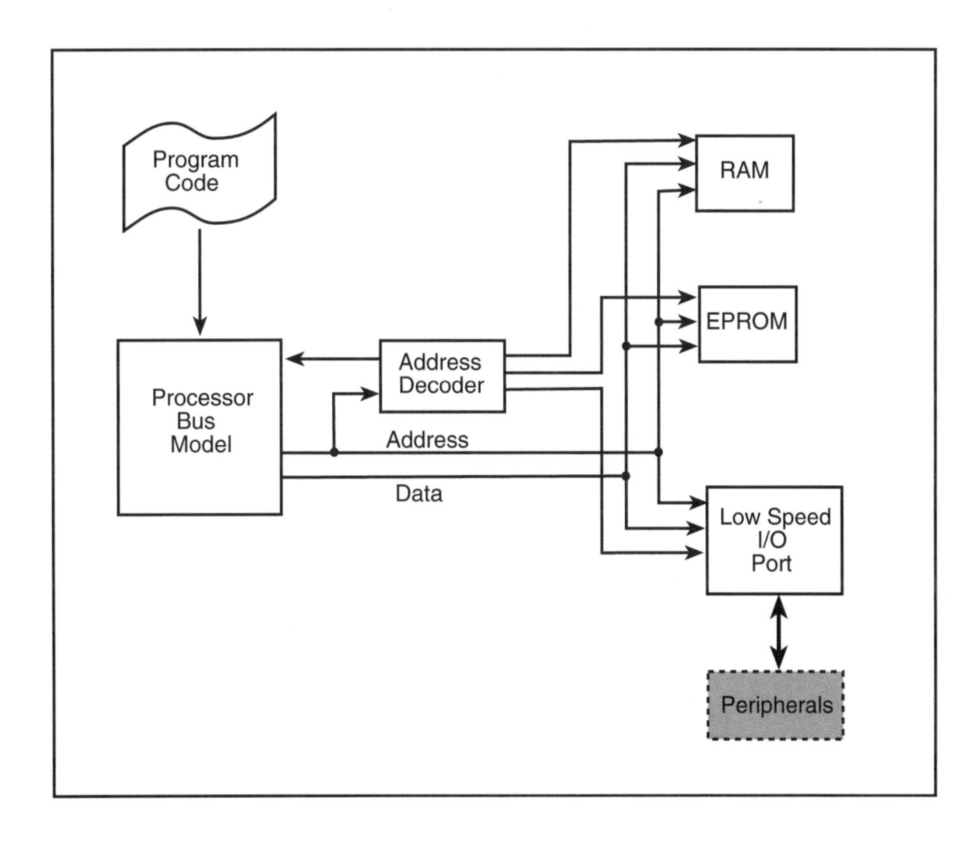

Fig. 3.5 Delays of Entire Modules Are Easy to Implement in an HDL

decode and low-speed I/O port will be synthesized and include any logic and flip-flops needed to perform their functions. Timing is important in the system simulation. At the RTL level, it is possible to see if the processor bus timing matches the RAM and EPROM timing. It can be determined if the decoder has too much delay or if the read/write timing of the I/O port meshes with the processor's requirements. The timing response of each block can be added to the model.

The read timing of the RAM is given in Figure 3.6. When the RTL model detects a read cycle, it can instantaneously get the data from its memory array and present it on the bus, but a fast response does not correspond to reality. The delay, shown in Figure 3.6 as Tvavd must be implemented in the model to reflect the time actually needed for the RAM to access and present valid data. The

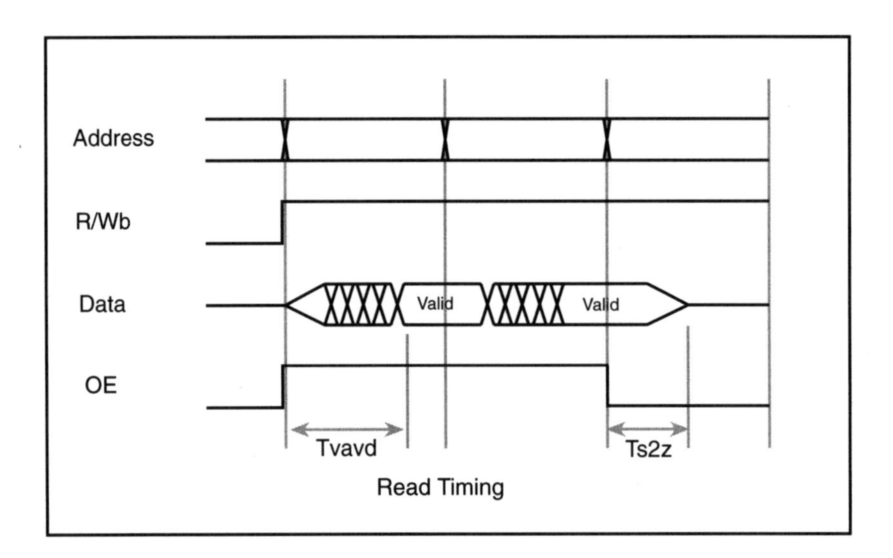

Fig. 3.6 RAM Timing Diagram

response time of the address decode cannot be instantaneous, but should reflect a delay based on the maximum delay it can have and still work in the system. The I/O port also needs bus timing to match the processor's characteristics. The processor model comes from the vendor with timing that matches the processor's real operation. The processor cycle time provides a check of the timing of all the other blocks. If a block meets the bus cycle time, it will work when fabricated.

A snippet of Verilog code, shown in Example 3.2, demonstrates how to implement the Tvavd and Ts2z delays in the memory model.

Example 3.2

```
1. 'define Tvavd   10   // data delay out of memory
2. 'define Ts2z     5   // delay of deselect to tristate
3. module RAM (addr, data, sel, rw);
4.    input [15:0]   addr;
5.    inout [15:0]   data;
6.    input       sel, rw;
7.    reg [15:0]   mem_array [0:65536], data_internal;
8.    // data bus tristate. Bi-directional.
9.    assign #Ts2z data = (sel) ? data_internal : 16'bz;
```

Example 3.2 (Continued)

```
10.    always @ (addr, rw)
11.    begin
12.       // read memory
13.       if ((rw === 1'b1) && (sel === 1'b1))
14.          #Tvavd data_internal = mem_array [addr[15:0]];
15.       // write memory
16.       if ((rw === 1'b0) && (sel === 1'b1))
17.          mem_array [addr[15:0]] = data;
18.    end
19. end module
```

Note when the memory is read, the assignment of the data from the array to the bus is delayed by the time Tvavd. The data bus response to the sel signal is also delayed by Ts2z. Whenever the timing of a module is known, it should be implemented in the RTL model; however, HDL languages offer different types of delays. It is important to understand how the delay is applied to ensure the model mirrors the real world. In Verilog, the two main default types are regular and intra-assignment. The effects of both types on continuous blocking, and nonblocking assignments are discussed below.

3.2.3 Delay with a Continuous Assignment Statement

The regular delay applied to the continuous assignment statement provides an inertial delay. An inertial delay means that the inputs must change and remain at their new values longer than the specified delay before the output is affected. A continuous assignment statement with a regular delay of 5 time units is shown in Example 3.3.

Example 3.3

```
Assign #5 sel = address15 | address16 | address17;
```

The output, sel, is simply the OR of the inputs address15, address16, and address17. Logically, whenever one of the address signals goes high, sel goes high; however, delay changes that fundamental assumption slightly. The relationship between the input and the output signals is shown in Figure 3.7.

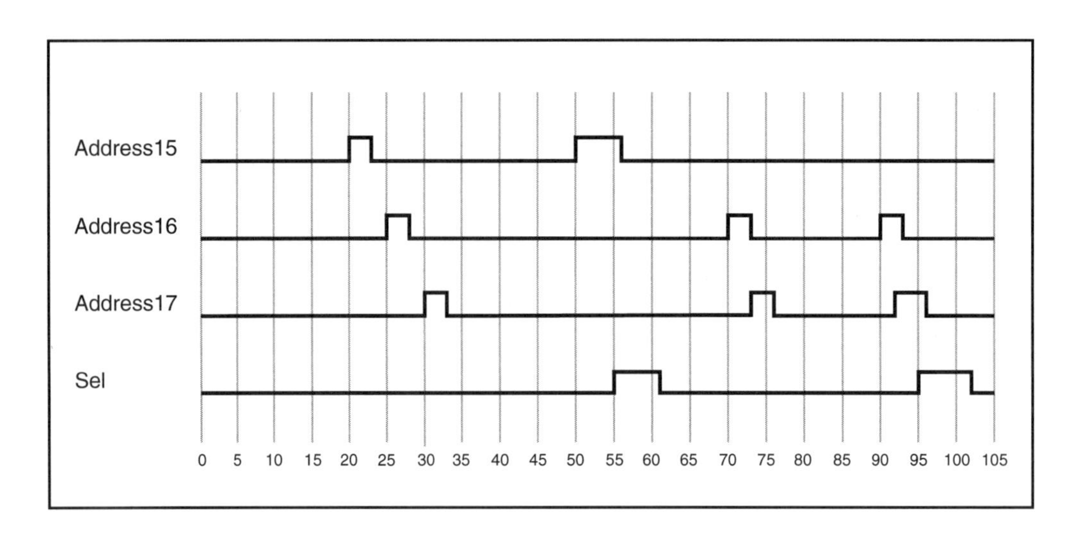

Fig. 3.7 Signals Corresponding to Example 3.3

At 20ns, each input sequentially goes high for 3ns. Each input stays high for less time than the specified delay of 5. The output does not change because the delay is inertial and no input is high-longer than the delay. At 50ns, address15 goes high for 6ns. After the input signal has been high for 5ns, the output responds and produces a pulse 6ns wide. At 70ns, both address16 and address17 go high for 3ns, but they are coincident and do not satisfy the inertial delay requirement, so the output does not change. At 90ns, a 3ns-wide pulse on address16 overlaps a 4ns-wide pulse from address17. The simulator interprets the overlap as meeting the delay requirement and a 7ns pulse occurs on the output.

Both the continuous assignment statement and the regular delay operate like combinatorial logic. Just as the delay through a gate suppresses glitches, so does the regular delay when used with a continuous assignment statement.

3.2.4 Delay in a Process Statement

Process statements, such as always or initial, support two types of assignment statements: blocking and nonblocking. The effects of regular and intra-assignment delays on both types of statements are shown below.

A regular delay with a blocking assignment is given in Example 3.4.

Example 3.4

```
always @(posedge clk)
begin
   #2 q1 = d;
   #2 q2 = d;
   #3 q3 = d;
end
```

A blocking statement means the simulator is blocked from moving on to any subsequent statement until the present one is complete. A regular delay delays evaluation of the inputs. The output signals that correspond to the process in Example 3.4 reveal exactly how a regular delay in a blocking statement works. Refer to Figure 3.8.

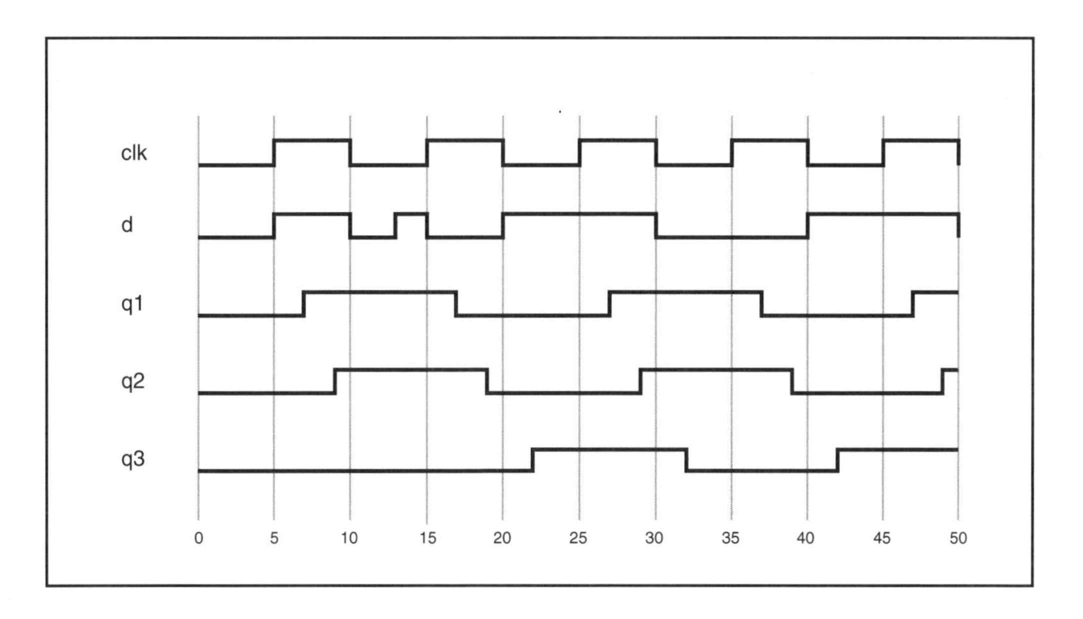

Fig. 3.8 Signals from the Blocking Assignment Statements with Regular Delays from Example 3.4

At 15ns, when the clock goes high, the simulator begins to execute the first statement, #2 q1 = d. It interprets it to mean: after a delay of 2 time units, assign the current value of d to q1. At 17ns, d is zero, so q1 becomes zero. The simulator waits at the first statement until it is completely finished; then it moves to the second statement. The second statement means the same as the first: wait 2 time units, then assign the present value of d to q2. Waiting an additional 2 time units means the value of d at 19ns is assigned to q2. At 19ns, d is zero, so q2 is assigned a zero. The simulator stays at the second line until the assignment to q2 takes place; then it moves to execute the third statement. The last statement has a delay of 3 time units. Like the previous regular delays, the simulator waits the specified time, 3 time units, then assigns d to q3. In this case, d changed to a one at 21ns, so when the simulator evaluates d at 22ns, it assigns a one to q3. The important concepts to remember about regular delays and blocking assignments are:

☞ **Blocking Assignments:** Finish executing the current, including the delay, before moving to the next line.

☞ **Regular Delays:** Wait the specified delay before evaluating the input signals and determining the output signal.

3.2.5 Intra-Assignment Delays

The intra-assignment delay is defined as follows.

☞ **Intra-Assignment Delay:** Upon execution, immediately evaluate the input signals and determine the value of the output signal. Wait the specified delay before assigning the value to the output.

The regular delay waits, evaluates, then assigns. The intra-assignment delay evaluates, waits, then assigns. An intra-assignment delay with blocking assignment statements is given in Example 3.5 along with the process statement that generates the input signal d.

Example 3.5

The waveforms in Figure 3.9 show how the input is evaluated immediately upon execution. At 15ns when the clock goes high, the first statement immediately grabs the value of d. The positive edge of clk triggers both the evaluation of d and its transition. At clk's positive edge, d has not yet changed and does not change until after it is grabbed by the q3 = #2 d assignment statement. As a result, the value assigned to q3 is d's value just before the clock's rising edge. At 15ns, d's value is one, so a one is grabbed and 2 time units later, at 17ns, a one is assigned to q3. The execution of the first statement is done, so the execution of the second assignment statement begins. At 17ns, the value of d is zero, so a zero value is grabbed by the second assignment statement and is assigned 2 time units later to q4.

```
always @(posedge clk)
begin
   q3 = #2 d;
   q4 = #2 d;
end
always @(posedge clk)
begin
   d <= ~d;
end
```

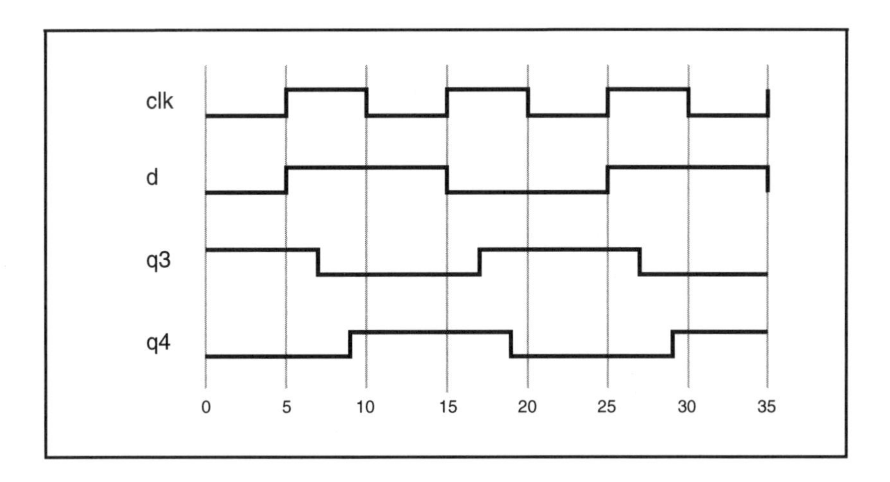

Fig. 3.9 Signals from Blocking Assignment Statements with Intra-Assignment Delays from Example 3.5

The operation of the intra-assignment delay is the same with nonblocking assignment statements, but the operation of a non-blocking statement does affect the output.

☞ **Nonblocking assignment:** Execute all nonblocking statements simultaneously. Do not execute them serially.

Nonblocking assignment statements with intra-assignment delays are given in Example 3.6.

The waveforms in Figure 3.10 show the value of d at the rising edge at 15ns to be a one. Two nanoseconds later, the value of one is assigned to both q7 and q8. Since the delay is the same in both statements, both outputs change at the same time.

Of the delay and assignment types described above, continuous assignment statements with regular delays closely model combinatorial logic. However, nonblocking assignment statements with intra-assignment delays in an always block, controlled by the clock, exactly model a flop-flop or sequential logic.

The regular and intra-assignment delays with continuous blocking and nonblocking assignment statements allow the designer to put delays anyplace in the circuit; however, assigning delays to possibly every line of RTL code takes a lot of time. As discussed in the section on synthesis, detailed timing should wait until synthesis or layout is complete. At the RTL level, it is sufficient to describe delays the boundaries and not lower. A higher level of granularity saves time developing code and also provides enough timing information to do meaningful analysis until the synthesis is complete. All HDL languages can express delays between module inputs and outputs. The approach taken in Verilog is presented below.

3.2.6 The Verilog Specify Block

Delays between module input and output ports in Verilog are described in a specify block. Delay is not the only timing parameter that can be expressed in the specify block, but all timing checks occur only between module input or ioput ports and output or ioput

Example 3.6

At the rising edge of clk, both assignment statements start execution. As shown in Figure 3.10, the positive edge of clk at 15ns causes both assignment statements to grab d's value. With blocking statements, the second statement was not executed until the first was completed, but with nonblocking statements both immediately start execution. Since the delay is intra-assignment, the input signal is immediately evaluated; then both statements wait 2 time units before assigning the evaluated result to the outputs.

```
always @(posedge clk)
begin
   q7 <= #2 d;
   q8 <= #2 d;
end
```

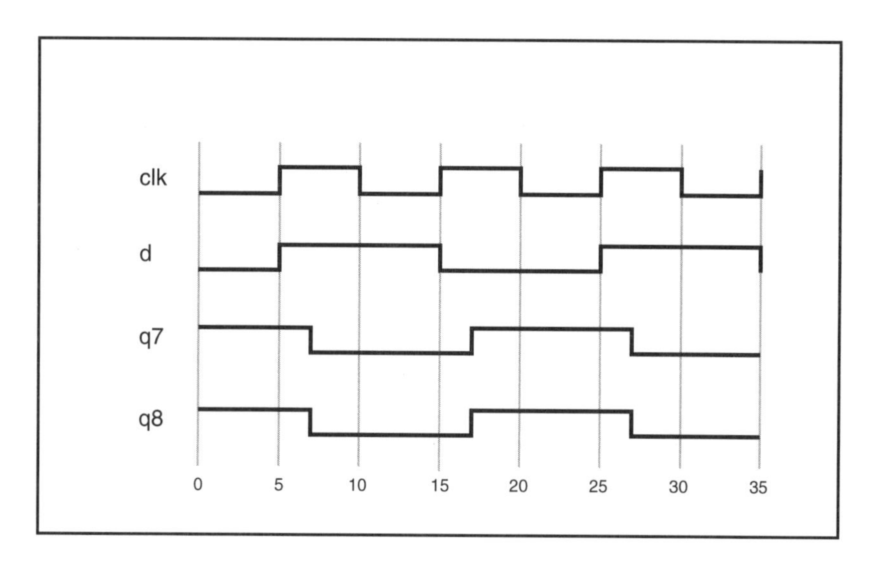

Fig. 3.10 Signals from Nonblocking Assignment Statements with Intra-Assignment Delays from Example 3.6

ports. A RAM memory module is again used to show how timing delays and verification are easily implemented in the specify block. A synchronous memory easily displays what types of checks can be done. The memory has the following timing requirements as shown in Table 3.1. Although numerous other parameters are needed to specify correct operation, these are sufficient to show how timing checks are defined. A diagram of the timing given in Table 3.1 is shown in Figure 3.11.

Table 3.1 Synchronous Memory Timing Parameters

Parameter	Time (ns)	Parameter	Time (ns)
Tclk_period	20	Tclk_data_valid	9
Tclk_high_min	9	T0_to_z	0.1
Tclk_low_min	7	Tz_to_1	0.3
Taddr_clk_setup	4	T1_to_z	0.1
Taddr_clk_hold	3	Tz_to_0	0.2
Tsel_clk_setup	4	Trise	0.5
Tsel_clk_hold	12	Tfall	0.3

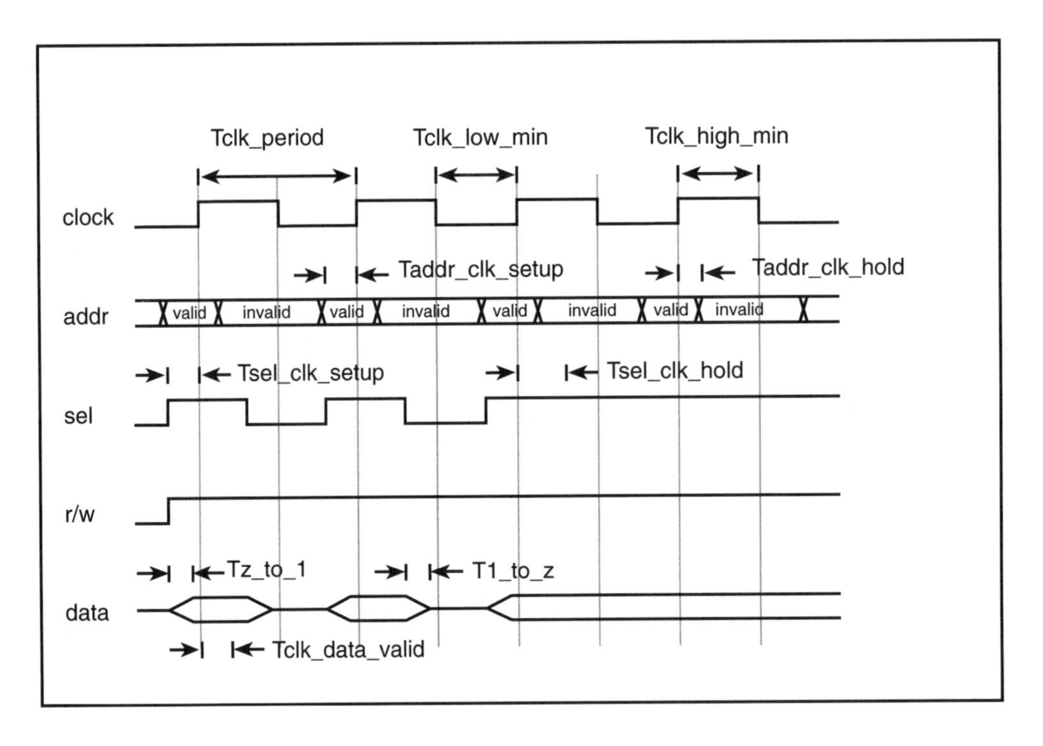

Fig. 3.11 Partial Timing Diagram for a Synchronous RAM

The RAM Verilog model is given below in Example 3.7.

Example 3.7

All the timing parameters listed in Table 3.1 are codified in the specify section. Each parameter is listed as a specparam. The parameter names of Table 3.1 directly correspond to the specparam names for easy correlation. The specparam statements span lines 30 through 43. The paths through the module are declared and described in terms of the timing parameters.

The memory model has only two paths with defined delays: clock to data and sel to data. The statement that defines the delay from the rising edge of the clock to valid data out is on lines 46 and 47 of Example 3.7. If the input signal, sel, is active, the delay from the rising edge of the clock to valid data out is defined in the parentheses following the equal sign. The delay from clock to valid data is Tclk_data_valid and the rise and fall times of internal signals are Trise and Tfall. The value for clk_data_valid is combined with the rise and fall times to provide more accurate delays.

```
 1. module RAM (clk, addr, data, sel, rw);
 2.
 3. input          clk;
 4. input [15:0]   addr;
 5. inout [15:0]   data;
 6. input          sel, rw;
 7.
 8. reg [15:0]     mem_array [0:65536], data_internal;
 9. reg            tprob;
10.
11. // data bus tri-state. Bi-directional.
12. assign data = (sel) ? data_internal : 16'bz;
13.
14. // Always statement that does the actual read and write
15. always @ (posedge clk)
16.    begin
17.    // read memory
18.    if ((rw === 1'b1) && (sel === 1'b1))
19.       data_internal = mem_array [addr[15:0]];
20.
21.    // write memory
22.    if ((rw === 1'b0) && (sel === 1'b1))
23.       mem_array [addr[15:0]] = data;
24.    end
25.
26. // The specify block where all the timing and verification is
       placed.
27. /******************************************************/
28. specify
```

Example 3.7 (Continued)

```
29. // define timing parameters
30.     specparam Tclk_period = 20;
31.     specparam Tclk_high_min = 9;
32.     specparam Tclk_low_min = 7;
33.     specparam Taddr_clk_setup = 4;
34.     specparam Taddr_clk_hold = 3;
35.     specparam Tsel_clk_setup = 4;
36.     specparam Tsel_clk_hold = 12;
37.     specparam Tclk_data_valid = 9;
38.     specparam Trise = 0.5;
39.     specparam Tfall = 0.3;
40.     specparam T0_to_z = 0.1
41.     specparam Tz_to_1 = 0.3
42.     specparam T1_to_z = 0.1
43.     specparam Tz_to_0 = 0.2
44.
45.     // declare module path and apply delay
46.     (if sel) (posedge clk *> data) = (Tclk_data_valid + Trise,
47.                               Tclk_data_valid + Tfall, 0, 0, 0, 0);
48.
49.     (negedge sel *> data) = (0, 0, T0_to_z, 0, T1_to_z, 0);
50.     (posedge sel *> data) = (0, 0, 0, Tz_to_1, 0, Tz_to_0);
51.
52. // do timing verification like set & hold, etc.
53.     $period (posedge clk, Tclk_period, tprob);
54.     $width (posedge clk, Tclk_high_min, 0, tprob);
55.     $width (negedge clk, Tclk_low_min, 0, tprob);
56.
57.     $setup (addr, posedge clk, Taddr_clk_setup, tprob);
58.     $hold (addr, posedge clk, Taddr_clk_hold, tprob);
59.
60.     $setuphold (sel, posedge clk, Tsel_clk_setup, Tsel_clk_hold,
        tprob);
61.
62. endspecify
63.
64.     // Report the time of every timing violation
65.     always @ (tprob)
66.     begin
67.        $display (%0d: "Timing violation found", $time);
68.     end
69.
70. end module
```

The meaning of the numbers in parentheses, lines 46 through 50, is summarized in Example 3.8. The first term defines the time it takes for a signal to transition from zero to one, the second is the time to transition from one to zero, the third is zero to high impedance, high impedance to one, one to high impedance, and high impedance to zero.

Example 3.8

```
(0 -> 1, 1 -> 0, 0 -> z, z -> 1, 1 -> z, z -> 0)
```

For the memory module, the delay from the positive edge of the clock to valid data only needs to have the zero to one and one to zero transitions defined because clock transitions do not cause the data bus to tristate.

The sel signal does cause the data bus to tristate, so the delay statements that define the relationship between the sel input and the data bus, lines 49 and 50 in Example 3.7, do not specify 0->1 or 1->0 delays. The statement on line 49 defines the time it takes to tristate the bus when sel goes inactive. Line 50 defines the time for the bus to leave tristate when sel becomes active.

Periods, pulse widths, and setup and hold times are also checked to see if they are in the specification. Lines 53 through 55 check the clock period, the time it is high and the time it is low. The setup and hold times of the address with respect to the clock's rising edge are checked in lines 57 and 58. The setup and hold times of sel to the rising edge of clock are checked in line 60. The formats of the verification statements are explained in Example 3.9. The notifier toggles every time a violation is found. The always statement after the specify block, lines 65 through 68, is activated when the notifier toggles to report the time of the violation.

Example 3.9

```
$period (ref_event, limit, notifier);
$width (ref_event, limit, threshold, notifier);
$setup (data_event, ref_event, limit, notifier);
$hold (ref_event, data_event, limit, notifier);
$setuphold (ref_event, data_event, s_limit, h_limit, notifier);
```

The specify block in Verilog HDL provides a convenient and powerful way to add timing to modules. It offers the right level of timing for RTL code. More specific and involved timing is available after synthesis or layout, automatically through the use of CAD tools. Do not spend time at the RTL level adding too much detail. Simply take advantage of any model-level timing offered by the simulator.

3.2.7 Timing in Gate-Level Code

HDL languages can simulate a design on the gate level where every gate is instantiated in the net list. Most designs do not start at the gate level but as RTL code and then go through synthesis to get gates. Gate-level simulations are important when the gate and interconnect delays are included because they provide insight into how the fabricated chip will work. Gate-level simulations are discussed in the synthesis section in the context of the standard delay format (SDF) file.

3.2.8 Synthesis and Timing Constraints

The object of synthesis is to produce a logically correct circuit from the RTL that meets the timing requirements. The synthesized logic gates should be correct by construction; however, the timing may not meet specification. The key to synthesis and obtaining correct timing is to provide reasonable timing constraints. Various approaches to I/O constraints, accounting for routing estimates, feeding timing information to a floorplanner or simulator, and synthesis strategies are discussed in this chapter. Any mention of the *synthesis tool* or *commands* refers to Synopsys Design Compiler and its related modules. All other synthesis tools have similar capabilities.

3.2.8.1 Synthesis Priorities A designer may have certain priorities when designing a circuit like low-power, high-speed, small-area, first-time manufacturability, etc. The synthesis tool also has priorities that may not coincide with the designer's goal. Synthesis walks a balance between what are called design rule constraints (DRC) and optimization constraints. Design rules are imposed on synthesis by the physical limitations of the technology library chosen to implement the design. Design rules deal with the following elements:

1. Maximum fanout per gate
2. Maximum transition time of a signal
3. Maximum allowable capacitance per net

The designer specifies optimization constraints to control these elements:

1. Speed

2. Area

3.2.9 Design Rule Constraints

As the synthesis tool translates the RTL into gates, it tries to meet the speed and area constraints requested by the designer. If the library is pushed to its limits and the tool must choose between meeting an optimization goal or a design rule priority, it satisfies the DRCs first. DRCs must take precedence over optimization constraints because if the gates of a library cannot meet the designer's requirements there is nothing that can be done except get a higher performance library (in terms of speed) or a library with smaller cells (in terms of area).

Although the library limitations play a role in forming the DRCs, the designer can also set limits on the library by specifying maximum fanout, transition, and capacitance to provide margin in the design. The designer must be sure that the DRCs are consistent for the entire design by propagating all user-set limits to all levels through appropriate use of .synopsys_dc.setup files. When setting DRCs, first consult the library to understand its limitations. Even if the designer chooses to use the same limits specified by the library, put them in the synopsys_dc.setup file that pertains to the design, so there are no questions what the limits are and where they are applied.

The following Design Compiler commands set DRC limits.

☞ **set_max_fanout:** Every input pin of every gate of the library has a fanout-load attribute. The sum of all fanout loads connected to an output cannot exceed the max_fanout limit. The command limits only the number of gates driven by any given output. Loading from wire capacitance is not controlled with this command.

- ☞ **set_max_transition:** The transition time is the amount of time it takes to charge or discharge a node. It is a product of the signal-line capacitance and resistance. The command set_max_transition watches the RC delay on a wire. In an effort to stay below the max_transition limit, the synthesis tool may increase the drive capacity of a gate to better swing the load or limit the capacitance and resistance by setting constraints that can be passed on to the floorplanner. The characteristics of the wire, such as area, capacitance, and resistance, are found in the wire-load model.
- ☞ **set_max_capacitance:** There are two components to a load on a net: fanout (other gates) and interconnect capacitance. The command set_max_capacitance checks to see that no gate drives more capacitance than the limit whether the source be interconnect or gate capacitance. There is no direct correlation between the command and net delay, simply between the command and capacitance. The wire-load model details the capacitance of a wire.

The three constraints mentioned above must be used in conjunction to ensure that the limits of the library are not exceeded. In the case where the library constraints do not match the limits set by the designer, the synthesis tool will meet the more restrictive value.

3.2.10 Optimization Constraints

After the DRCs are met, the synthesis tool works on optimizing the design. The most important optimization constraint is speed. The synthesis tool uses an internal static timing analyzer to determine if a path meets the required time. Static timing analysis (STA) is described more fully in the next section; however, in a nutshell, it sums up the delays of every element in a path to see if the total delay is faster or slower than required. The delay is measured from one sequential element to the next. A sequential element is considered to be a flip-flop or a latch. A more precise definition of a path is from an output pin to an input pin with a setup-and-hold-time requirement. Synthesis and STA work best on synchronous designs. There are techniques to deal with asynchronous circuits; however, if it is possible to design the circuit to be synchronous, it will fit into the modern ASIC flow with fewer exceptions that need to be manually checked.

There is another design practice, in addition to synchronous design, that enhances the use of synthesis and STA. The static timing analyzer in the synthesis tool considers the clock tree to be ideal which means there is no delay between the clock source and the input of any gate. In a design where the clock signal goes directly from the clock tree to the gates, its operation is nearly ideal. Any design technique, such as gated clocks, that places delays in the clock's path will not work unless the amount of delay in the clock is quantified. It is possible to use the clock skew parameter to account for the delay in the clock, but it must include both the skew of the tree and the delay through gates. The clock delay through gates is not automatically measured, so it may be a difficult figure to arrive at. It is a good design practice to not gate the clock.

The designer can control the synthesized speed of the circuit with commands explained below.

- ☞ **create_clock:** At a minimum, the synthesis tool must know the clock's period and duty cycle. The clock sets the time allowed for signals to propagate between sequential elements. The create_clock command also specifies clock skew.
- ☞ **set_input_delay:** The delay of the input of a module is assumed to be zero. The circuit, shown in Figure 3.12, has four inputs and is considered a module. Two inputs go directly to flip-flops while the other two inputs go through gates before they reach a flip-flop. The delay time for input a to reach the flip-flop is input_delay. If the set_input_delay command specifies the input_delay as 2ns, then the synthesis tool measures the delay of a and b as 2ns and if necessary modifies the design appropriately to still work at speed. The delay of input c or d is: input_delay + and-gate delay + or-gate delay. The value of input_delay is added to the gate delays to arrive at the final

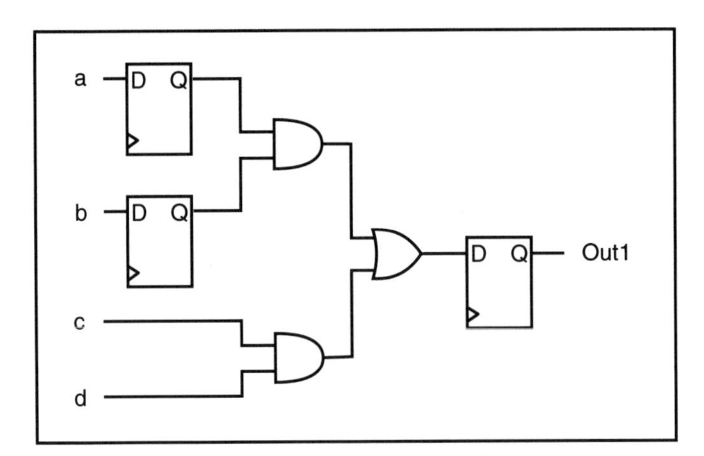

Fig. 3.12 The set_input_delay Command Adds Additional Delay to Module Input Times

speed of the path. If there is a lot of input_delay, the synthesis tool chooses faster gates to maintain the overall speed specified by the designer.

☞ **set_output_delay:** The delay out of a module can be increased by the amount specified by set_output_delay. The module shown in Figure 3.13 has two output signals. The delay of

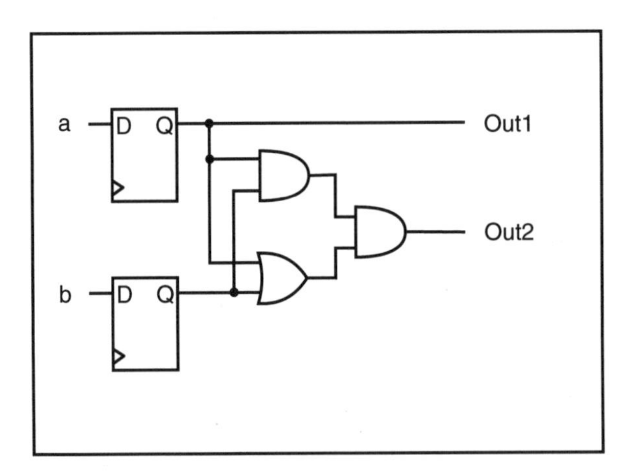

Fig. 3.13 The set_output_delay Command Adds Additional Delay to Module Output Times

out1 is: flip-flop propagation delay + output_delay. If the set_output_delay command sets out_delay to 5ns, the delay of out1 is 5ns longer than the propagation delay of a flip-flop. The delay of out2 is: flip-flop propagation delay + [maximum of (and-gate or or-gate delay)] + and-gate delay + output_delay. Once again, the output_delay adds to the circuit's inherent delays.

- ☞ **set_max_delay:** Timing constraints can be placed on asynchronous paths with set_max_delay and set_min_delay. The values set by these two commands determine the time allowed to propagate through a path not controlled by a clock.
- ☞ **set_min_delay:** Refer to set_max_delay.
- ☞ **set_max_area:** The area constraint is set by a single command. If an area is specified, the synthesis tool will try to keep the area of both the gates and the wires under the max_area limit. The area of the wires can only be estimated if it is specified in the wire-load model.

Once the design rule and optimization constraints are specified, the synthesis tool works to find the correct gates to implement the logic functions specified in the RTL code with the timing specified by the designer. Timing in ASIC standard cell circuits cannot be fully understood without knowing the source of gate and wire delays.

3.2.11 Gate and Wire-Load Models

As gates are chosen by the synthesis tool to implement the logic functions described in the RTL, the synthesis tool uses an internal static timing analyzer to add up gate and wire delays to see if their total stays within the timing constraints. The delays are completely dependent on the process technology of the library chosen for fabrication. The vendor provides the delay numbers for both gate and wire-load models.

3.2.11.1 Gate Models Every gate available to the synthesis tool must be described as a library model. A sample of a library that contains only an AND-gate is given in Example 3.10. The description of the AND2 cell provides all the information the synthesis tool needs to determine if it can meet timing and area requirements. The area of the cell is given on line 3. Each input pin, lines 6 and 11, is described with its associated capacitance, lines 8 and 13, so the synthesis tool can calculate total fanout loads for the driving gates. The output is described in terms of the logic function it performs, line 18, in addition to the response of the output with respect to each input. The timing response of the output with respect to input A is given in lines 20 through 27, and with respect to input B in lines 28 through 36. The most important timing figure is the propagation delay for rising and falling transitions as controlled by each input, which is given in lines 20 and 21 and 29 and 30. The output rise and fall times and slopes are given along with the output resistance.

Example 3.10

```
 1. library (proc_35) {
 2.     date: "September 29, 2001"
 3.     revision: 1.9
 4. cell(AND2) {
 5.     area: 3
 6.     pin(A) {
 7.     Direction: input
 8.     Capacitance: 1.2
 9.     fanout_load: 1.0
10. }
11. pin(B) {
12.     Direction: input
13.     Capacitance: 1.2
14.     fanout_load: 1.0
15. }
16. pin(Z) {
17.     Direction: output
18.     Function: "AB"
19.     Timing(): {
20.         intrinsic_rise: 1.38
21.         intrinsic_fall: 0.97
22.         rise_resistance: 1.00
23.         fall_resistance: 1.00
24.         slope_rise: 1.00
25.         slope_fall: 1.00
26.         related_pin: "A"
27.     }
28.     Timing(): {
29.         intrinsic_rise: 1.38
30.         intrinsic_fall: 0.97
31.         rise_resistance: 1.00
32.         fall_resistance: 1.00
33.         slope_rise: 1.00
```

Example 3.10 (Continued)

```
34.        slope_fall: 1.00
35.        related_pin: "B"
36.        }
37.    }
38.    }
39. }
```

3.2.11.2 Wire-Load Models The synthesis tool estimates wire delays using a wire-load model that relates a net's estimated length to estimated capacitance and resistance. The manual calculation of the characteristics of a line is fully described in section 3.3. The synthesis tool uses the same techniques to find the RC delay of each net. There is a statistical aspect of the wire delay calculation. The actual length of each net is unknown to the synthesis tool; however, it makes a guess using statistics of routing from the reference design. Based on the statistical estimate of length, it calculates area, capacitance, and resistance. The delays determined using vendor wire-load models are inaccurate because the model is design dependent. If your design is not similar to the reference design used to make the wire-load model, there is significant error; however, the estimated delay decreases the number of synthesis iterations because estimated delay is better than ignoring it altogether.

Fortunately, more accurate wire-load models can be generated specifically for a given design. As soon as the RTL code is complete, the design can be synthesized and given to a floorplanner, then a place-and-route tool. The information from the preliminary route is fed back into the synthesis tool to make custom wire-load models that are much more accurate than the vendor-supplied models because they are design specific. The most accurate wire-load models are available after the place-and-route procedure once each wire's exact dimensions are known. A wire-load model is shown in Example 3.11.

Example 3.11

```
Wire-load ("16x16") {

Resistance : 0.1 ;
Capacitance : 1.85;
```

Example 3.11 (Continued)

```
Area : 1.4;
Slope: 1.0;
Fanout_length (1, 1.6);
Fanout_length (2, 2.9);
Fanout_length (3, 5.5);
Fanout_length (4, 10.1);
}
```

Another approach to compensating for inaccurate wire-load models is to synthesize to a faster clock than the design will actually use. The synthesis tool chooses gates capable of driving larger loads, so when the accurate delays are fed back to the simulator after layout, the extra speed is used up in driving the lines. Another technique is to overestimate the capacitance values of the gates in the library so the synthesis tool chooses gates with extra drive capacity. The problem with any approach based on deliberate overdesign is that the area is larger than it may have to be and the amount of overcompensation, whether it be in time or capacitance, is merely a guess. The best approach is to have the most accurate models possible, which for wire-load models means that the data from an early floorplan should be used to develop accurate wire-load models.

3.2.12 The Synthesis Flow

By now it is clear that synthesis is a key step to obtaining the correct timing in ASIC standard cell design, but it is not the only step. Synthesis selects the gates used to implement the logic functions, but they are fashioned into the final form for fabrication by a floorplanner, and place-and-route tools. The process of converting RTL code into final layout is an iterative process. The major steps are listed below with emphasis on how the major tools interact.

1. Synthesize using vendor library and statistical wire-load models.
2. Write out timing constraint information (SDF) from the synthesis tool to be used by the floorplanner.
3. Using a timing driven floorplanner, plan the overall placement that meets the timing constraints from synthesis.

4. Place and route the design.
5. Extract cluster values (PDEF), delay values (SDF), and parasitic estimated values (RC). Feed the information to the floorplanner.
6. Create wire-load models using the information from place and route.
7. Back annotate the wire-load models into the synthesis tool (or a stand-alone static timing analyzer). Analyze the design to see if it meets timing requirements.
8. If the timing is close, use the reoptimize_design command to fix the few problems that exist. Generate new constraint information; then go to step 3 when done.
9. If the timing is not close, use the new wire-load models to synthesize again. Generate new constraints and return to step 3.
10. If the timing has plenty of slack, do the final place and route. Go to step 9.
11. If the timing is perfect after the final floorplan place-and-route iteration, the design is done. Otherwise fix the few minor problems that exist with the in-place optimize option of the synthesis tool.
12. Write out final delay and parasitic values for use in a static timing analyzer or in RTL-gate simulations as a final verification that the correct timing was achieved.

There is a tremendous amount of communication between the synthesis tool, the floorplanner, the place-and-route tool, static timing analyzers, and even the RTL simulator. The information sent from each tool helps the next tool in the process do its job better. Each iteration brings the design closer to the correct timing which is verified with either a static timing analyzer or RTL-gate simulations with full-timing back annotation.

Three common file formats pass the information between the tools: physical data exchange format (PDEF), standard delay format (SDF), and resistance/load scripts. Each is described below.

☞ **Physical Data Exchange Format (PDEF):** The PDEF file contains information about the clustering of cells. The synthesis tool determines which cells should be close to each other (in a cluster) based on how the RTL file is organized. Since most designers partition their designs based on logic functions, the synthesis tool also groups logically. Once the floorplanner gets the netlist, it places the cells together based on timing or routing considerations. It generates a PDEF file based on physical placement that may not be anything like the logical groupings generated by the synthesis tool. A typical PDEF file is shown in Example 3.12.

Example 3.12

```
(CLUSTERFILE
(PDEFVERSION "2.0")
   (DESIGN "top")
   (DATE "October 29, 2001")
   (VENDOR "Intrinsix")
   (DIVIDER /)
   (CLUSTER
   (NAME "MultB1")
   (X_BOUNDS   0.0   150.0)
   (Y_BOUNDS   0.0   163.0)
      (NAME "MultSub1")
      (X_BOUNDS   0.0   25.6)
      (Y_BOUNDS   0.0   89.3)
      (CELL (NAME U24/U78)   (LOC   3.8    16)
      (CELL (NAME U24/U45)   (LOC   16.0   46.5)
      (NAME "MultSub2")
      (X_BOUNDS   25.6   78.8)
      (Y_BOUNDS   0.0   89.3)
      (CELL (NAME U55/U14)   (LOC   30.0   42.7)
      (CELL (NAME U55/U83)   (LOC   51.0   46.5)
   )
   (NAME "MultB2")
   (X_BOUNDS   150   204.0)
   (Y_BOUNDS   0.0   79.0)
      (NAME "Nts1")
      (X_BOUNDS   154.2   36.0)
      (Y_BOUNDS   185.0   0.0)
      (CELL (NAME U47/U64)   (LOC   153.8   24.0)
      (CELL (NAME U86/U37)   (LOC   167.0   28.2)
   )
)
```

☞ **Standard Delay Format (SDF):** The standard delay format file specifies delays. It is a case-sensitive format. The synthesis tool uses the SDF file to pass timing constraints to the floorplanner, an action known as forward-annotation. It uses the PATHCONSTRAINT parameter to tell the floorplanner the amount of propagation delay allowed for critical paths. The format of the PATHCONSTRAINT statement is given in Example 3.13.

Example 3.13

```
(PATHCONSTRAINT port_start [intermediate_node, ...] port_end (rise
      time) (fall time))
```

A simple SDF constraint file for the circuit shown in Figure 3.14 is shown in Example 3.14. The three highlighted paths are described.

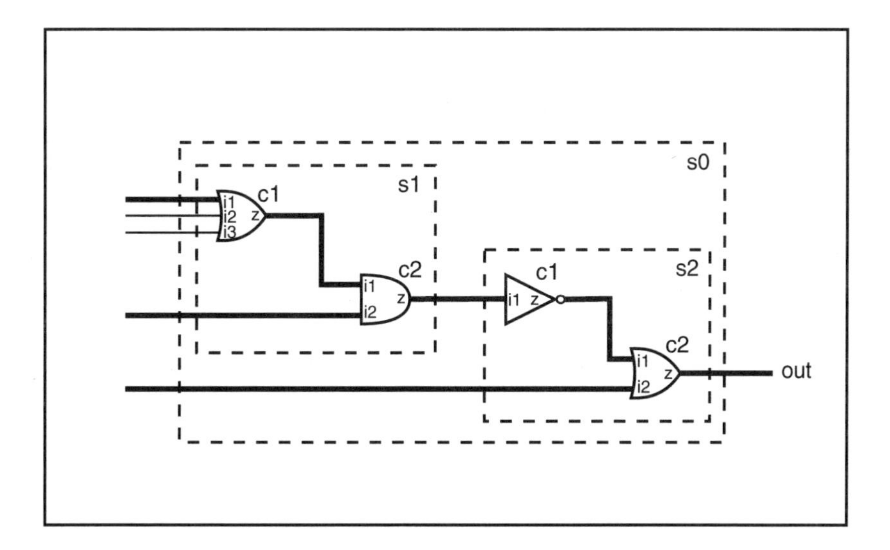

Fig. 3.14 Circuit Corresponding to the SDF Constraint File of Example 3.14

Example 3.14

```
   (DELAYFILE
// Start of the sdf header. This file contains all typical data
   (SDFVERSION "1.0")
   (DESIGN "test")
```

Example 3.14 (Continued)

```
   (DATE "Monday January 30 08:30:33 PST 1999")
   (VENDOR "Intrinsix Corp.")
   (PROGRAM "delay_find_forward")
   (VERSION "3.6")
   (DIVIDER /)
   (VOLTAGE 5.0:5.0:5.0)
   (PROCESS "typical")
   (TEMPERATURE 85:85:85)
   (TIMESCALE 1ns)
// The intrinsic delays of each cell used in the design. Equivalent
       to gate delays.
(CELL
   (CELLTYPE "test")
   (INSTANCE s0)
   (TIMINGCHECK
   (PATHCONSTRAINT   s1/c1/i1                        // start node
            s1/c1/z                                  // intermediate nodes
            s1/c2/z
            s2/c1/z
            s2/c2/z                                       // end node
            (3.76:3.76:3.76) (3.44:3.44:3.44))            // times
   (PATHCONSTRAINT   s1/c2/i2                             // start node
            s2/c1/z
            s2/c2/z                                       // end node
            (2.65:2.65:2.65) (2.52:2.52:2.52))            // times
   (PATHCONSTRAINT   s2/c2/i2                             // start node
            s2/c2/z                                       // end node
            (1.54:1.54:1.54) (1.33:1.33:1.33))            // times
     ))
```

The same file format passes delay information from the synthesis tool to the RTL simulator and from the floorplanner/router to synthesis or RTL. The format can define the delays across a module, gates, or interconnect. Timing for setup, hold, setuphold, skew, width, and period are also valid parameters. Delays can also be specified to be absolute or incremental. Most HDL simulators use a subset of the SDF parameters. The designer does not need to do anything with the SDF file. The simulator accepts and assigns the delays using built-in system tasks. For Verilog, the command to read an SDF file is $sdf_annotate. The user can specify if minimum, typical, or maximum timing values are extracted from the SDF file and can set a scale factor if desirable.

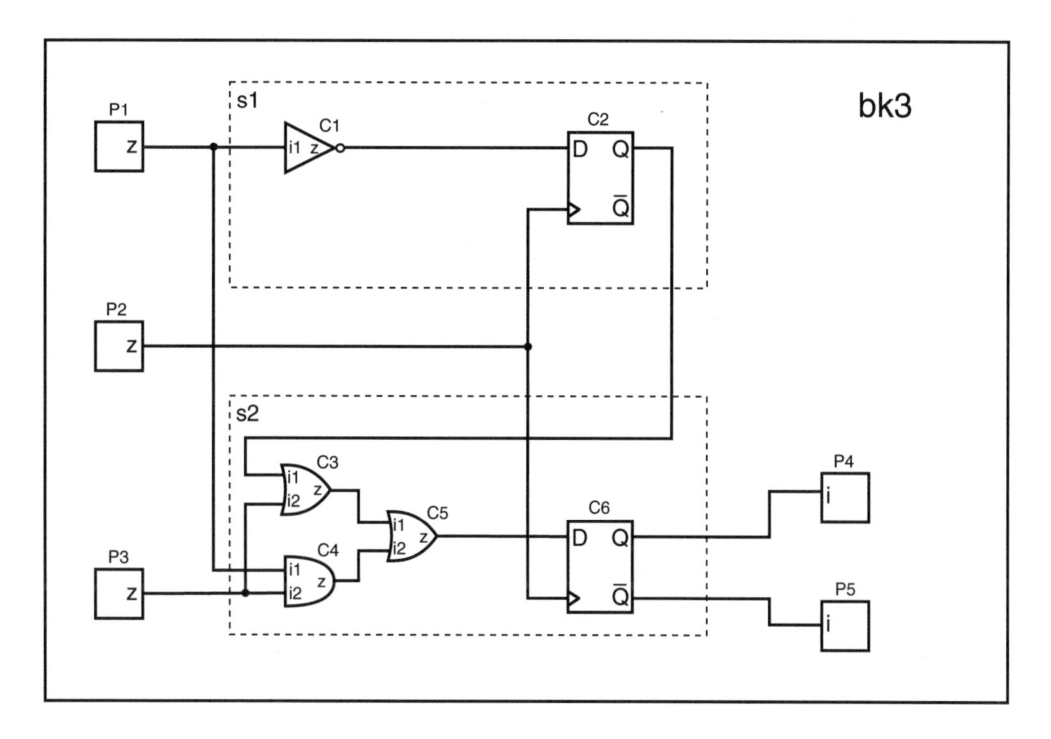

Fig. 3.15 Circuit Corresponding to the SDF File of Example 3.15

The SDF file for the circuit shown in Figure 3.15 is given in Example 3.15. It includes the most common parameters used by RTL simulators. Each construct is also described.

Example 3.15

```
 1. (DELAYFILE
 2.     // Start of the sdf header.
 3. // This file contains all typical data.
 4.     (SDFVERSION "1.0")
 5.     (DESIGN "test")
 6.     (DATE "Monday January 30 08:30:33 PST 1999")
 7.     (VENDOR "Intrinsix Corp.")
 8.     (PROGRAM "delay_find")
 9.     (VERSION "3.6")
10.     (DIVIDER /)
11.     (VOLTAGE 5.0:5.0:5.0)
12.     (PROCESS "typical")
13.     (TEMPERATURE 85:85:85)
14.     (TIMESCALE 1ns)
15. // description of interconnect delays.
```

Example 3.15 (Continued)

```
16. (CELL
17.      (CELLTYPE "test")
18.    (INSTANCE bk3)
19.    (DELAY
20.    (ABSOLUTE
21.    (INTERCONNECT P1/z     s1/c1/i   (.163:.163:.163)
                                     (.147:.147:.147))
22.    (INTERCONNECT P1/z     s2/c4/i2  (.152:.152:.152)
                                     (.139:.139:.139))
23.    (INTERCONNECT P2/z     s1/c2/clk (.102:.102:.102)
                                     (.099:.099:.099))
24.    (INTERCONNECT P2/z     s2/c6/clk (.109:.109:.109)
                                     (.101:.101:.101))
25.    (INTERCONNECT P3/z     s2/c3/i2  (.178:.178:.178)
                                     (.165:.165:.165))
26.    (INTERCONNECT P3/z     s2/c4/i1  (.176:.176:.176)
                                     (.163:.163:.163))
27.    (INTERCONNECT s1/c1/z  s1/c2/d   (.184:.184:.184)
                                     (.175:.175:.175))
28.    (INTERCONNECT s1/c2/q  s2/c3/i1  (.171:.171:.171)
                                     (.163:.163:.163))
29.    (INTERCONNECT s2/c3/z  s2/c5/i1  (.185:.185:.185)
                                     (.173:.173:.173))
30.    (INTERCONNECT s2/c4/z  s2/c5/i2  (.146:.146:.146)
                                     (.137:.137:.137))
31.    (INTERCONNECT s2/c5/z  s2/c6/d   (.189:.189:.189)
                                     (.176:.176:.176))
32.    (INTERCONNECT s2/c6/q  P4/i      (.169:.169:.169)
                                     (.155:.155:.155))
33.    (INTERCONNECT s2/c6/qn P5/i      (.187:.187:.187)
                                     (.174:.174:.174))
34.    )))
35. // The intrinsic delays of each cell used in the design.
       Equivalent to gate delays.
36. (CELL
37.    (CELLTYPE "INV")
38.    (INSTANCE s1/c1)
39.    (DELAY
40.    (ABSOLUTE
41.    (IOPATH   i   z   (.323:.323:.323) (.311:.311:.311))
42.    )))
43. (CELL
44.    (CELLTYPE "DFF")
45.    (INSTANCE s1/c2)
46.    (DELAY
47.    (ABSOLUTE
48.    (IOPATH   clk   q   (.417:.417:.417) (.404:.404:.404))
49.    ))
50.    (TIMINGCHECK
51.       (SETUP D (posedge clk) (.260))
52.       (HOLD D (posedge clk) (.000))
53.       (WIDTH (negedge clk) (1.60))
54.       (WIDTH (posedge clk) (1.73))
55.    ))
56. (CELL
```

Example 3.15 (Continued)

```
57.     (CELLTYPE "DFF1")
58.     (INSTANCE s2/c6)
59.     (DELAY
60.     (ABSOLUTE
61.     (IOPATH   clk   q    (.379:.379:.379) (.367:.367:.367))
62.     (IOPATH   clk   qn   (.421:.421:.421) (.414:.414:.414))
63.     ))
64.     (TIMINGCHECK
65.        (SETUP D (posedge clk) (.302))
66.        (HOLD D (posedge clk) (.000))
67.        (WIDTH (negedge clk) (1.64))
68.        (WIDTH (posedge clk) (1.79))
69.     ))
70.  (CELL
71.     (CELLTYPE "OR2")
72.     (INSTANCE s2/c3)
73.     (DELAY
74.     (ABSOLUTE
75.     (IOPATH   i1   z   (.304:.304:.304) (.298:.298:.298))
76.     (IOPATH   i2   z   (.304:.304:.304) (.298:.298:.298))
77.     )))
78.  (CELL
79.     (CELLTYPE "OR2")
80.     (INSTANCE s2/c5)
81.     (DELAY
82.     (ABSOLUTE
83.     (IOPATH   i1   z   (.304:.304:.304) (.298:.298:.298))
84.     (IOPATH   i2   z   (.304:.304:.304) (.298:.298:.298))
85.     )))
86.  (CELL
87.     (CELLTYPE "AND2")
88.     (INSTANCE s2/c4)
89.     (DELAY
90.     (ABSOLUTE
91.     (IOPATH   i1   z   (.337:.337:.337) (.325:.325:.325))
92.     (IOPATH   i2   z   (.337:.337:.337) (.325:.325:.325))
93.  )))
```

The delay times are specified as triplets. The first group of three numbers is the rise time. The second is the fall time. The three numbers, separated by colons, are supposed to represent (minimum:typical:maximum) delays; however, many tools write only one case at a time. In the SDF file example above, all three numbers in the parentheses are the same. The header specifies that they are the typical case. Completely new files would need to be written for the minimum and maximum cases.

The time scale, line 13, is also given in the header as nanoseconds. For this process, most of the gates have a propagation delay of between 0.250 and 0.45ns. In all cases for this example, the delay times are declared as absolute. If the delays had been incremental, the final delay would be the sum of a base time and the increment specified in the SDF file. In this case, there are no base time and no increments. The timing number states how long it takes for a signal to propagate through a line or gate.

3.2.12.1 INTERCONNECT The interconnect command is used to specify interconnect delay. After synthesis, but before routing, the interconnect delays are estimates based on wire-load models. After place and route, the exact dimensions of each line are extracted and the RC delay calculated. The interconnect statement on line 20 describes the delay from the pad input P1 to the input of the inverter C1. The signal propagation delay is 0.163ns for a rising edge and 0.147ns for a falling edge. The format for the statement is:

```
(INTERCONNECT port_start port_end (rise times) (fall times))
```

3.2.12.2 IO PATH The ioport statement specifies input to output cell delays. The path can be from any input/ioput port to any legal output/ioput port. Lines 35 through 41 describe the propagation delay through the inverter C1. The propagation delay of a rising edge is 0.323ns and a falling edge is 0.311ns. The format for the statement is:

```
(IOPATH input_port output_port (rise times) (fall times))
```

3.2.12.3 SETUP, HOLD, SETUPHOLD, WIDTH, PERIOD These statements are timing checks of sequential devices like flip-flop or latches. The timing checks must be specified in the TIMINGCHECK part of the cell definition. The purpose of each timing check matches its name. The definition of the D flip-flop (DFF), lines 42 through 54, describes the clk to q propagation delay, on line 47, as 0.417ns for a rising edge and 0.404ns for a falling edge. It also provides timing checks for the data setup and hold times on lines 50 and 51. Minimum pulse widths are also specified for the clock on lines 52 and 53. Line 52 requires the clock to be low for at least 1.6ns. The minimum clock high time of 1.73ns is given on line 53. The formats for the timing check statements are:

```
(TIMINGCHECK
   (SETUP data_signal reference_signal (time))
   (HOLD data_signal reference_signal (time))
   (WIDTH (posedge/negedge signal) (time))
   (SETUPHOLD data_signal reference_signal (setup time) (hold time))
   (PERIOD (posedge/negedge signal) (period))
)
```

3.2.12.4 Resistance/Load Scripts The floorplanner sends line resistance and load scripts to the synthesis tool to make new wire-load models. The format is shown in Example 3.16.

Example 3.16

```
Set_resistance 0.1569 "aout"
Set_load 0.673 "aout"
```

3.2.13 Synthesis Tips

Some general suggestions can help the synthesis process converge to the correct timing faster.

1. **Are the constraints realistic?** Understand the limitations of the cells in the technology library even before starting RTL coding. Make a short table of common gates with propagation delays for various loads. Does the RTL code expect too much out of the gates? Does pipelining need to be used to make the target speed realistic? Test the synthesis process on small sections of the code to make sure it can meet the speed expectations before the entire design is done and it is discovered that the library is not up to the challenge.
2. **Have false paths and multicycle paths been identified?** A false path is a path that does not need to meet timing requirements. Signals that activate test modes are good examples of false paths. A multicycle path is one that is not expected to complete its computation in a single cycle. Identification of both types of paths eliminates unnecessary warnings from the tools.

3. **Have minimum and maximum delays been set for all asynchronous paths?** As mentioned earlier, synchronous designs are better for the synthesis design methodology. If a path simply cannot be made synchronous, be sure that both minimum and maximum delays are specified to set boundaries on its operation.
4. **Do the timing violations require a change in architecture?** Are the timing violations so egregious they cannot be solved with minor fixes or more iterations through the design cycle? Look at timing violations with an eye on architecture. Always ask if an architectural change would make many timing problems go away.
5. **Can minor timing problems be solved with incremental compiles?** At times, the synthesis, floorplan, place, and route cycle can be like the golpher game at the arcade—after hitting one golpher over the head with the bat, another three pop up. Sometimes doing another synthesis run to solve a minor problem can result in problems in several other paths. When the timing is close, use the incremental compile option to iron out any remaining problems.

3.2.14 Back Annotation to Gate-Level RTL

Delays can be applied to the RTL model at two important stages in the design process. After synthesis, it is possible to add precise fanout and gate delays to the RTL model along with estimated interconnect delays. Once a design is through floorplanning, placement, and routing, accurate interconnect delays can be added to the gate and fanout delays known from synthesis. Gate-level timing can be tested at either point: after synthesis or layout. Usually the vendor sign-off procedure requires verification with postlayout, back-annotated timing. Two timing verification approaches are common: static timing analysis and back-annotated gate-level simulations.

3.3 POSTLAYOUT TIMING

The most accurate simulations of a device are those that use delays extracted from the actual layout. Synthesis provides accurate fanout figures, but only estimated propagation delays for signals. Once the layout is complete, the exact dimensions of every line are known and can be used to determine the RC loading and propagation delay.

Even extracted delays are only estimates of the delays of the fabricated device; however, at the postlayout state, the accuracy of the model is sufficient to ensure first-time functional silicon. Two generally accepted methods for verifying the postlayout model have been previously discussed: static timing analysis and gate-level RTL simulations with back-annotation from an SDF file. The delays extracted from layout provide more accuracy than any prelayout estimates; however, there are a few instances where manual verification of the extracted delay merits attention.

3.3.1 Manual Line-Propagation Delay Calculations

Interconnect can be modeled at several different levels each with decreasing amounts of accuracy. Complete 3D modeling is the most accurate, followed by 2.5D, transmission lines, and finally lumped RC analysis [Chiprout98]. Except at ultrahigh frequencies of 1GHz or above, transmission lines and lumped components provide sufficient accuracy. Most standard cell methodologies do not require any manual intervention to correctly determine interconnect delays. There are some situations, however, when it may be necessary. Generally these conditions do not occur in a standard cell methodology. If there is any deviation from the standard design flow sanctioned by the vendors or if the design is known to push the process limits, you may want to check for the following conditions:

1. Signal lines in polysilicon
2. Extremely long signal lines

Such cases will be rare; however, if they exist, it is worth the effort to manually verify the timing of a few lines. If the delay from the manual calculation is no more accurate than the delay already reported, there is no need to individually model any more lines.

There are extraction tools capable of determining the width and length of any line. If such tools are available, use them to get the information necessary to calculate the line resistance and capacitance as shown below. Except at ultrahigh frequencies, internal routing lines do not exhibit inductive characteristics, so inductance can safely be ignored in the model.

3.3.2 Signal-Line Capacitance Calculation

There are two components to line capacitance: plate and fringe. Capacitors form between a conductor and whatever lies underneath, whether it be the substrate or another conductor. Two conductors and the capacitors formed are shown in Figure 3.16. Capacitors form where the metal-2 line crosses over the metal-1 trace and where both traces cross over the substrate. The two types of capacitors that affect a trace are shown in Figure 3.17. A plate capacitor forms between the bottom of the trace and the underlying layer. A fringe capacitor exists between the sides of the conductor and whatever is underneath.

Total line capacitance is the sum of the capacitance of the bottom and both edges:

$$\text{Cline} = \text{Cplate} + 2\text{Cfringe} \qquad \textbf{(Eq. 3.1)}$$

Plate capacitance is calculated as follows:

$$\text{Cplate} = \frac{\varepsilon WL}{D} \qquad \textbf{(Eq. 3.2)}$$

where

ε is the dielectric constant of the insulator (SiO2) between the trace and the substrate. A value for permittivity of free space, $\varepsilon 0$, is 8.85e-6 pf/um. The permittivity of SiO2 is 3.9*$\varepsilon 0$ or 3.45e-5 pf/um.

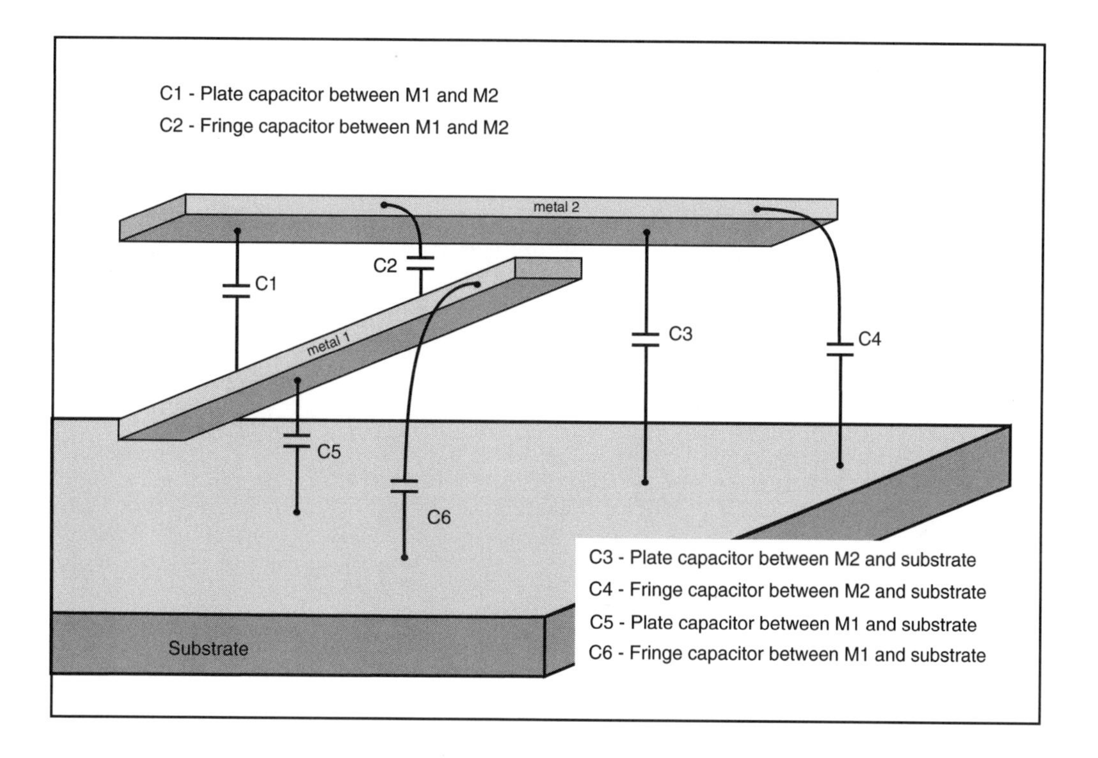

Fig. 3.16 Parasitic Capacitors Exist Between All Layers of a Process

The fringe capacitance is more difficult to model, but a close approximation is to convert each edge of the trace to a half cylinder then combine them into a separate, complete cylinder. Figure 3.18 diagrams the conversion. The cylinder represents the capacitance contributed by both edges and replaces the 2Cfringe term in equation 3.1. The formula to calculate the capacitance of a cylindrical conductor is:

$$\text{Ccylinder} = \frac{2\pi\varepsilon d}{\ln(1 + \{2D/d\}\{1 + \text{sqrt}[1 + (d/D)]\})} \quad \textbf{(Eq. 3.3)}$$

where

d = diameter of cylinder

D = distance from the substrate

ε = dielectric constant of SiO2

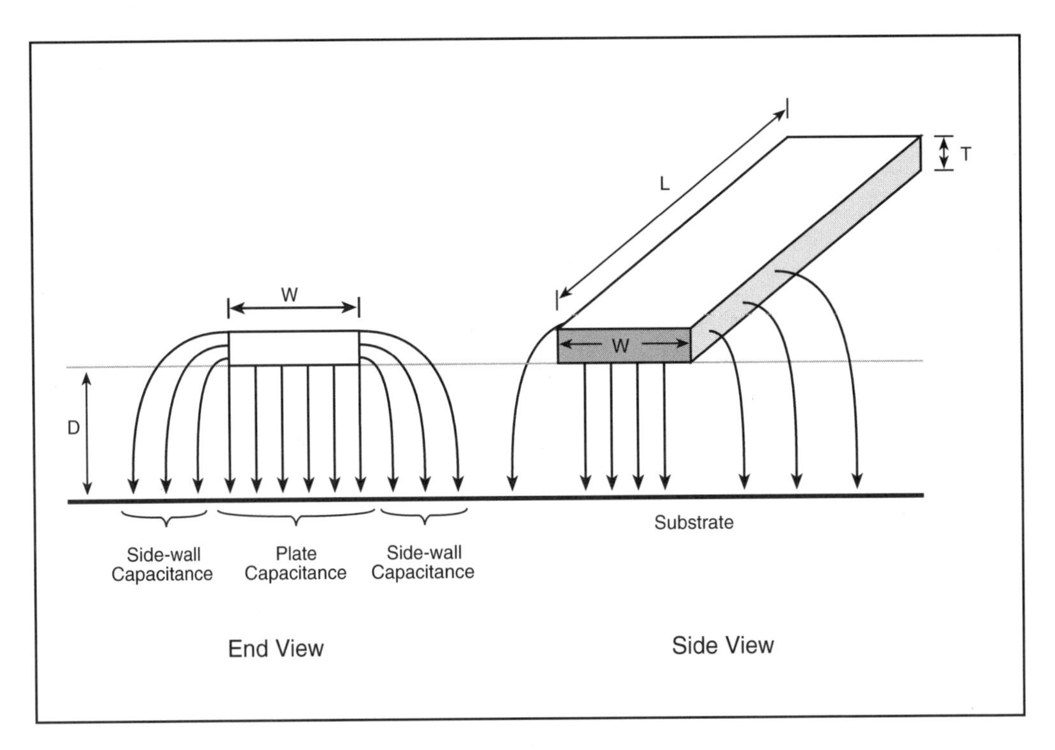

Fig. 3.17 Conductors Have Both Plate and Fringe Capacitors

Notice in Figure 3.18, the width of the rectangle is decreased by T/2 for the calculation of the plate capacitance. It would seem that its width should decrease by T since a T/2 slice was taken off each side; however, the cylinder represents all the edge capacitance plus a little bit of plate capacitance, so the rectangle width is reduced by a lesser amount. Refer to [GD85] for more in-depth information.

Many design rule specifications include values for both plate and fringe capacitance which are expressed as pf/um^2 and pf/um. If the values are not already available, calculate the plate capacitance as pf/um^2 and fringe capacitance as pf/um as follows:

$$Cp = \frac{\varepsilon}{D} \quad \textbf{(Eq. 3.4)}$$

$$Cc = \frac{2\pi\varepsilon}{\ln(1 + \{2D/T\}\{1 + sqrt[1 + (T/D)]\})} \quad \textbf{(Eq. 3.5)}$$

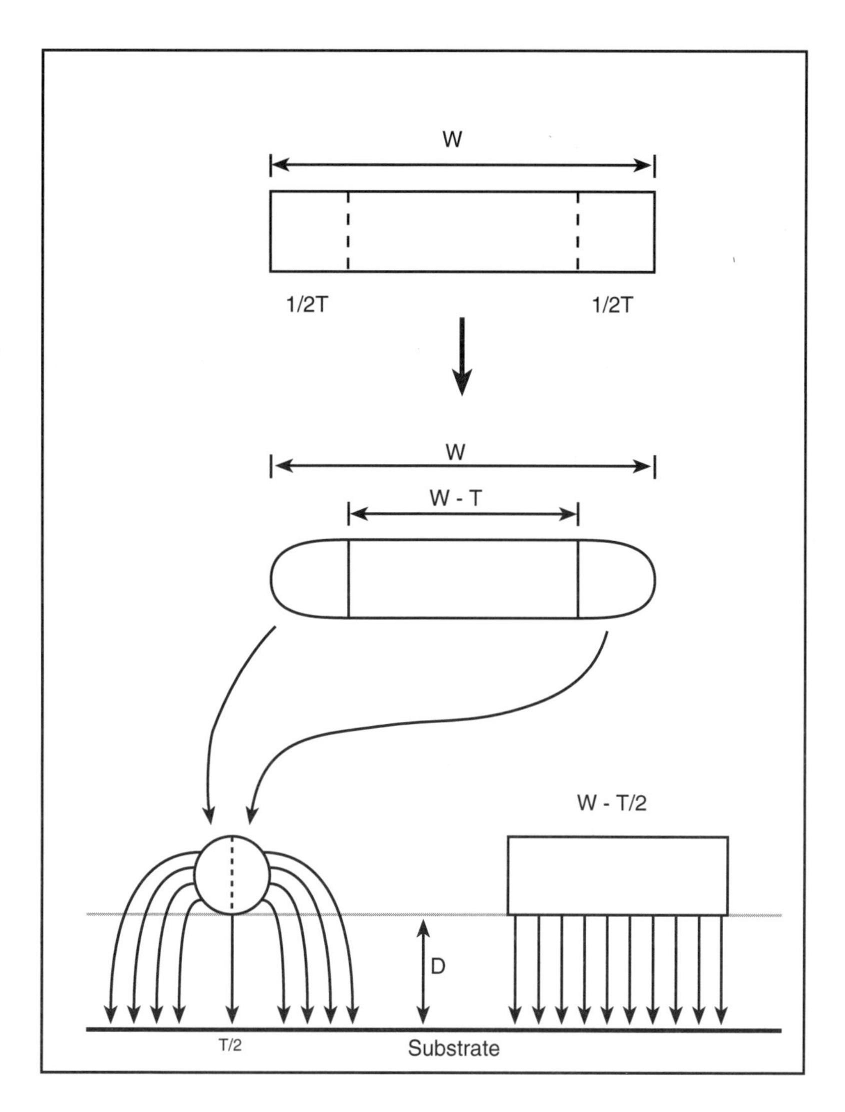

Fig. 3.18 Fringe Capacitance Can Be Modeled as a Cylindrical Conductor

The final equation to approximate the capacitance of a trace is:

$$Cline = CpWL + CcL \quad \textbf{(Eq. 3.6)}$$

Minimum line widths and typical trace thickness are listed in Table 3.2 for an 0.8um process. Typical capacitance values for a triple-layer metal process are given in Table 3.3. The dielectric constant for SiO2 is also given in Equation 3.7:

$$\varepsilon SiO2 = 3.45e{-5}\ pf/um \quad \textbf{(Eq. 3.7)}$$

The capacitances of several long lines for all of the above capacitance combinations are given in Table 3.4 below. The width is assumed to be the minimum width for the material listed.

Table 3.2 Typical Metal Widths for 0.8um Process

	T (um)	Minimum Width (um)
Poly	0.3	0.8
Metal 1	0.4	1.4
Metal 2	0.8	1.4
Metal 3	1.0	1.6

Table 3.3 Typical Capacitance Values for 0.8um Process

	D (um)	Cp (e-4 pf/um^2)	Cc (e-4 pf/ um)
Poly to Substrate	0.47	0.734	1.03
Metal 1 to Poly	0.52	0.664	1.11
Metal 1 to Substrate	0.97	0.356	0.884
Metal 1 to Diffusion	0.52	0.664	1.11
Metal 2 to Metal 1	0.45	0.767	1.56
Metal 2 to Poly	1.05	0.329	1.11
Metal 2 to Substrate	1.37	0.252	1.0
Metal 2 to Diffusion	1.05	0.329	1.11
Metal 3 to Metal 2	0.45	0.767	1.72
Metal 3 to Metal 1	1.3	0.265	1.11
Metal 3 to Poly	1.9	0.182	0.963
Metal 3 to Substrate	2.22	0.155	0.911
Metal 3 to Diffusion	1.9	0.182	0.963

Table 3.4 Capacitance of Long Lines

	Line Length (um)			
Total Line Capacitance (pf)	**500**	**1500**	**3000**	**6000**
Poly to Substrate	0.076	0.227	0.453	0.906
Metal 1 to Poly	0.095	0.286	0.572	1.14
Metal 1 to Substrate	0.066	0.197	0.393	0.787
Metal 1 to Diffusion	0.095	0.286	0.572	1.14
Metal 2 to Metal 1	0.117	0.35	0.699	1.4
Metal 2 to Poly	0.072	0.215	0.43	0.86
Metal 2 to Substrate	0.063	0.188	0.376	0.751
Metal 2 to Diffusion	0.072	0.215	0.43	0.86
Metal 3 to Metal 2	0.128	0.385	0.77	1.54
Metal 3 to Metal 1	0.07	0.21	0.42	0.841
Metal 3 to Poly	0.058	0.174	0.349	0.698
Metal 3 to Substrate	0.054	0.162	0.325	0.649
Metal 3 to Diffusion	0.058	0.174	0.349	0.698

Table 3.4 shows the capacitance of a line over a single material. For example, in the metal-2-to-substrate capacitance, it is assumed that there are no poly or metal-1 lines between the substrate and the metal 2. The amount of capacitance depends on the amount of area overlap and which layers are involved. Assuming that a line runs over a single material is not an accurate assumption. Routing between cells leaves myriad lines crossing one another, over transistors and the substrate. A good extraction tool can determine exactly what lies under a line; however, if the line crosses lots of other layers, making a capacitance model can be difficult. If the tool can report only the width and length of a trace, but not the layers underneath, the capacitance value for the trace to the substrate may have to be used. If a quick visual check reveals the line is over a different layer some part of the time, the model accuracy can be improved.

For example, if the line is a very long metal-2 line with half of it over metal 1, use the metal-2-to-metal-1 capacitance value for 50% of the trace's area and the metal-2-to-substrate value for the rest.

Also note that the capacitance scales linearly with the line length. A table, like Table 3.4, which contains relatively few entries, can be used to determine the capacitance of a line of any length by extrapolation.

3.3.3 Signal Line Resistance Calculation

The resistance of a line is measured in units called squares. The lines shown in Figure 3.19 have width W and length L. The length of each line is subdivided into squares W wide and W long. There are ten squares total for the line in 3.19a. Note that the squares are

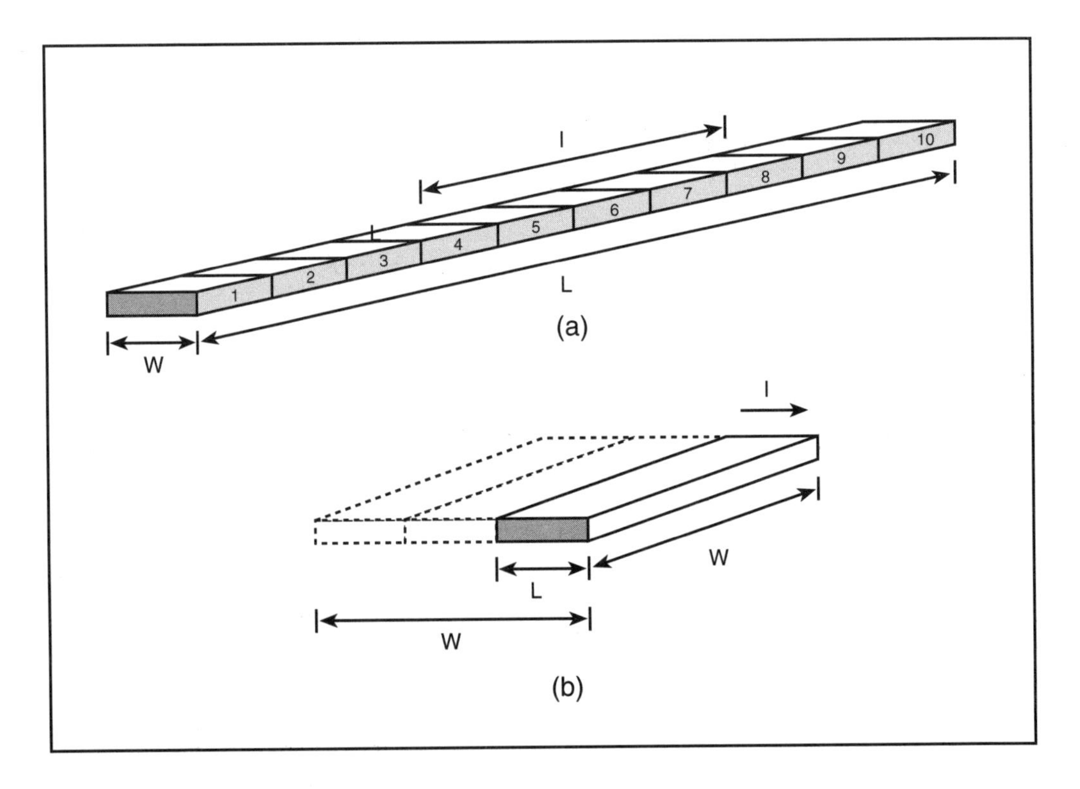

Fig. 3.19 Line Resistance Is Measured in Squares

marked off along the direction of the flow of current. The line shown in 3.19b has 0.33 squares.

Each type of material used in semiconductors has a characteristic resistance per square. The resistance of a line is calculated by finding the total number of squares and multiplying by the ohms per square value. Resistance values for each layer are given in the process design rules. Typical resistance values for a 0.8um process are given in Table 3.5.

The resistances of the lines drawn in Figure 3.19 for different materials are given in Table 3.6.

A look at the resistance of the different materials reveals why lines laid out in polysilicon need additional attention. The capacitance of poly to substrate is about the same as the capacitance between metal 2 and metal 3 or metal 1 and metal 2; however, its resistance is 160 times greater than the resistance of metal 1 and 400 times that of metal 2 and metal 3. Lines of polysilicon have a high RC delay. If the CAD tools do not account for its RC characteristics, the simulation would be highly inaccurate. Fortunately, all

Table 3.5 Typical Material Resistance in a 0.8um Process

	Ohms/square
Polysilicon	40
Metal 1	0.25
Metal 2	0.1
Metal 3	0.1

Table 3.6 Line resistances for Figure 3.19

	3.19a (ohms)	3.19b (ohms)
Polysilicon	400	13.3
Metal 1	2.5	0.083
Metal 2	1.0	0.033
Metal 3	1.0	0.033

routing in a standard cell process is done in metal. If any hand edits are made or if there are any deviations from a previously proven flow, check for poly routing. Special cells, that are more custom in nature, such as memory decoders, select lines in memories, or decoded buses may use poly as a routing layer to achieve higher density. If the timing of drop-in cells is not provided to include in the RTL model as described in section 3.2, do not neglect the RC delay of lines when characterizing the block's timing.

3.3.4 Signal Trace RC Delay Evaluation

Once the resistance and capacitance for a trace are known, the propagation delay can be determined. The easiest approach is to combine all the line resistance into a single resistor and all the capacitance into a single capacitor. Elementary circuit theory predicts the rise time from the 10% to the 90% voltage points of the lumped RC model to be 2.2RC; however, simulations of a distributed RC line show much less delay. Breaking the line into at least three, and not more than ten, equal sections, as shown in Figure 3.20, provides a model that is

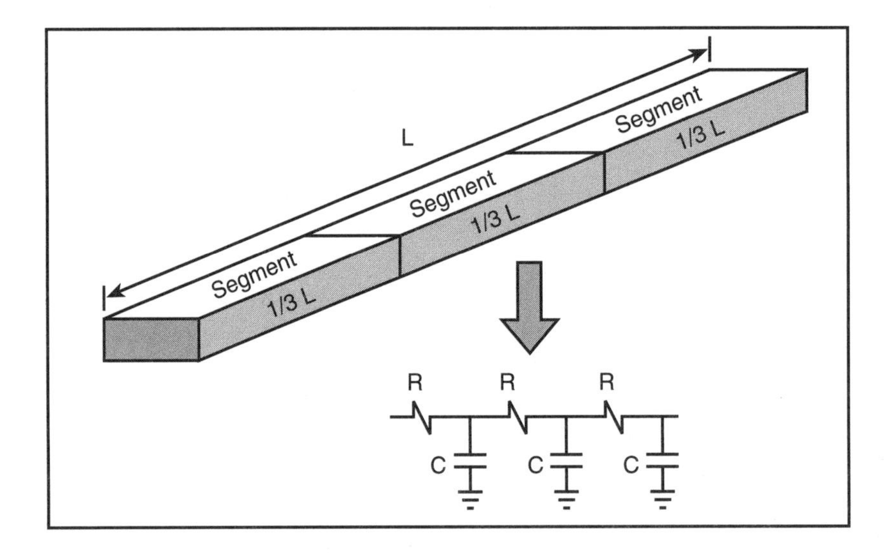

Fig. 3.20 Distributed RC Model Provides Greater Accuracy than Lumped Model

accurate to within 3% of a transmission line model. Simulations show that the 10%-90% delay of the distributed line approaches the RC time constant as the number of RC segments in the simulation increase [Wilnai71] and [Chiprout98].

The RC delay for lines of minimum width and various lengths are listed in Table 3.7. The capacitance in each case is that of the line to the substrate.

Polysilicon is the highest resistance routing material, so it results in the highest delays. Very few lines, if any, will be run in polysilicon, but those that are need a closer look. The delays are easily reduced by increasing the line width which in turn decreases the resistance. Although the capacitance also increases, the significant decrease in resistance results in lower propagation delays. Note that the delays in metal are minimal. They may be a bit higher if the line runs over closer layers like poly or other metal lines; however, for the most part the delay of a metal trace will not affect performance significantly. If the circuit is designed right on the edge of the process capabilities, long metal lines may need closer scrutiny. If there are several long lines on the device, simulate the distributed RC model of one of them to see if all need to be modeled more accurately.

3.4 ASIC SIGN-OFF CHECKLIST

Standard cell ASIC design is a partnership between the designer and the vendor who fabricates the device. The designer usually

Table 3.7 RC Delay Times

RC Time Constant	Line Length (um)			
(ns)	500	1500	3000	6000
Polysilicon	1.9	17	68	272
Metal 1	0.006	0.053	0.21	0.84
Metal 2	0.005	0.02	0.081	0.32
Metal 3	0.002	0.015	0.061	0.24

takes care of what is known as the front-end part of the design which consists of the functional specification through to postsynthesis gate-level simulations. The vendor's responsibilities are to supply the technology library and to perform the layout. Once the layout is done, the vendor provides the designer with an extracted SDF file for final verification. If the extracted design passes final verification, there is a high probability that the design will work when fabricated.

A checklist details which tasks are to be done by whom and how to ensure, or verify, that the task has been done properly. Most checklists cover at least the following items.

3.4.1 Library Development

The vendor is responsible for development of the standard cell library. The library must be characterized so accurate gate and interconnect models can be used in simulation and STA. One consideration when selecting a library is the variety of gates it offers. The richer the selection of gates, the more flexibility it offers the synthesis tool. Variety refers to the different flavors of any given cell such as low-power, high-speed, small area, etc.

3.4.2 Functional Specification

Only the designer can write the functional specification. Only the designer knows what the application requires. The specification may be determined in large part by components of a library if there are special cells or intellectual property (IP) that meets the application needs.

3.4.3 RTL Coding

The designer develops the RTL code that implements the functional specification. Some RTL code is much easier to synthesize into stable, workable circuits. If the vendor has any coding guidelines that help the synthesis tool better utilize the library, they should be followed.

3.4.4 Simulations of RTL

The designer is responsible to verify that the RTL code correctly implements the functional specification. Simulations at the RTL level should be thorough because this is really the only place where correct function can efficiently be verified. Simulations at the gate level are much too slow to be complete and STA does not verify function, only timing.

3.4.5 Logic Synthesis

Synthesis is the responsibility of the designer; however, the vendor may have to supply information about the library as the synthesis tool tries to make decisions about which cells to use and how to estimate the wire-loads. The synthesis tool generates both forward and backward annotation files. The forward annotation provides constraints to timing-driven layout tools while the back-annotated files provide delay information to either a simulator, for gate-level simulations, or a static timing analyzer.

3.4.6 Test Insertion and ATPG

The insertion of flip-flops for partial or full scan is done by the designer as is automatic test program generation (ATPG).

3.4.7 Postsynthesis Gate-Level Simulation or Static Timing Analysis

The designer is responsible for verifying the synthesized gates for functional correctness and for estimated performance. Whether the verification is done with a simulator or a static timing analyzer, the wire-loads are only estimates. The gate delays come from the technology library and are accurate. The delays are provided from the synthesis tool via a standard delay format (SDF) file.

3.4.8 Floorplanning

Some designers have the tools and capabilities to perform their own floorplanning. Most do not and the task falls to the vendor. Floorplanning takes information from synthesis to group the cells to meet the timing performance. It feeds back more accurate wire-load models to the synthesis tool and it provides the framework for place and route.

3.4.9 Place and Route

Most vendors drive the tools that actually place the cells and run the wires between them. The end result is the actual layout that, if it is verified to work properly, is used to make masks for fabrication. The vendor extracts gate and interconnect delays to be used in final verification. The clock tree is also developed at the same time by the vendor. Some tools exist to measure the load on the tree and develop the branches in such a way to minimize skew. The vendor performs simulations on the extracted clock tree to ensure that it meets specification.

3.4.10 Final Verification of the Extracted Netlist

The designer is responsible for final verification of the extracted netlist. The extracted netlist offers the most accurate model of the device. The exact fanout of every gate is known along with the delay of every signal line. The method of final verification, whether it be by simulation or STA, is determined by the vendor. Some vendors do not accept STA alone as proof that the device will work and require extensive back annotated simulations. Other vendors use STA as the only final verification tool. Most designers use both: complete STA analysis plus a few strategic simulation vectors.

3.4.11 Mask Generation and Fabrication

Once final verification is complete, the vendor assumes the responsibility of everything connected with fabricating the device.

3.4.12 Testing

Once the device is fabricated, the vendor does some limited tests to ensure the process met specifications. Production uses the vectors developed by the designer.

C H A P T E R 4

Programmable Logic Based Design

4.1 Introduction

Complex programmable logic devices (CPLDs) and field programmable gate arrays (FPGAs) are off-the-shelf parts that a user can program to quickly create an ASIC-like device. Compared to ASICs, they provide a faster and less risky path to the development of complex functionality. However, performance and cost may, to a varying degree, be sacrificed. These parts vary greatly in capability and are supplied in a large variety of programming technologies, architectures, sizes, packages, and speed grades by a host of vendors. Additionally, a variety of software development systems are provided by the programmable logic vendors themselves and by third-party CAE companies. Unfortunately, there is little commonality of architectures and tools between vendors. Because of this diversity, it becomes somewhat of a challenge for the CPLD or FPGA designer to evaluate which part best suits his requirements.

The ability to achieve a particular level of performance using a specific device to implement a design is influenced by the following factors:

- Process
- Functionality
- Architecture
- Tools

Process and architecture are important factors, but so too is the functionality that is to be implemented. A particular process provides an upper boundary on performance. For example, if a programmable logic family employs 0.5 micron CMOS, no matter what the designer does, only a level of performance that is supported by this technology will be achieved. In another dimension, the programmable logic device's structures characterized by the device's architecture provide the receptacles into which a design must be placed. How well the design fits into these structures affects performance. The complexity of the design functionality also affects the level of performance. Design functionality is entered into the system via schematic capture or hardware description language (HDL) code creation and synthesis. It is important to minimize the complexity of the functionality because extra levels of logic quickly add large delays in programmable logic devices. Unfortunately, it is very easy to create unintended complexity through the use of HDL code and synthesis. Therefore the designer must know what each HDL-code-construct-to-gate synthesis transform will be. The software tools used throughout the design process also have an end effect on performance. The quality of the tools is important, but so is the designer's ability to understand and use the tools properly.

It is the purpose of the material presented in this chapter to provide an overview of programmable logic design topics with an emphasis on timing-related issues. This chapter is divided into the following sections:

- **Programmable Logic Structures:** This section presents an introduction to the two types of programmable logic devices, the CPLD and the FPGA, and the components that make up these devices.
- **Design Flow:** The various parts of the programmable logic device design flow are outlined in this section.
- **Timing Parameters:** Timing parameters are defined, and then timing-derating factors and speed grading are discussed in this section.

- **Timing Analysis:** This section contains detailed descriptions of vendor-specific architectures along with their timing models. Several examples of how to use the timing models are presented.
- **HDL Synthesis:** A brief overview of HDL synthesis as applied to programmable logic design is presented in this section.
- **Software Development Systems:** A general overview of software development systems is given along with specific descriptions of the timing-related software from Actel, Altera, and Xilinx.

4.2 PROGRAMMABLE LOGIC STRUCTURES

CPLD devices are made up of PAL-like function blocks, while FPGA devices feature a gate-array-like architecture. Each of these device types contains the following key architectural elements:

- Logic blocks
- I/O blocks
- Routing resources
- Clock networks

The designer uses these elements to create a design that contains the desired functionality and meets the required performance criteria. As a signal passes into, around, and out of a device, each of these elements adds its contribution to the delay of the signal. After a design has been captured (via schematic entry or HDL-code creation), vendor-specific software, tailored to a device's specific programming technology and architecture, is used to configure the logic and I/O blocks, and control the interconnection of the programmable routing structures. The implementation of the design created by these tools, in conjunction with the timing parameters of the elements within the logic and I/O blocks, the routing structures, and the clock networks, then determines the resultant timing performance of the device.

Both CPLD and FPGA devices consist of a core of logic blocks surrounded by I/O blocks, all of which is interspersed by routing resources. The differences between the two lies mostly in the nature of the logic blocks and the interconnect structures. The structure of a typical CPLD is shown in Figure 4.1, and a very simplified view of an FPGA device is shown in Figure 4.2. CPLD architectures generally evolved as extensions to PAL architectures and their internal logic blocks tend to resemble the structure of these devices. FPGAs, on the other hand, started out with gate-array-like architectures having configurable logic elements sitting in a sea of configurable routing elements. Over time the architectures of the two device types have evolved such that in some cases they exhibit similar

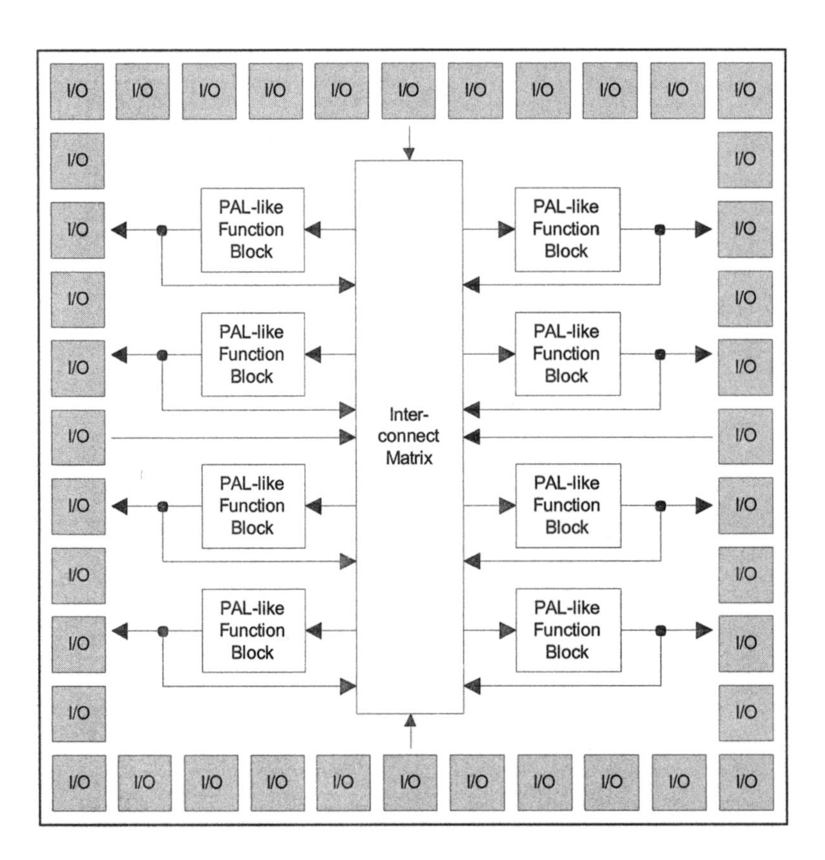

Fig. 4.1 CPLD Structure [Xilinx, 1998] Courtesy of Xilinx, Inc. Copyright © Xilinx, Inc., 1999. All rights reserved.

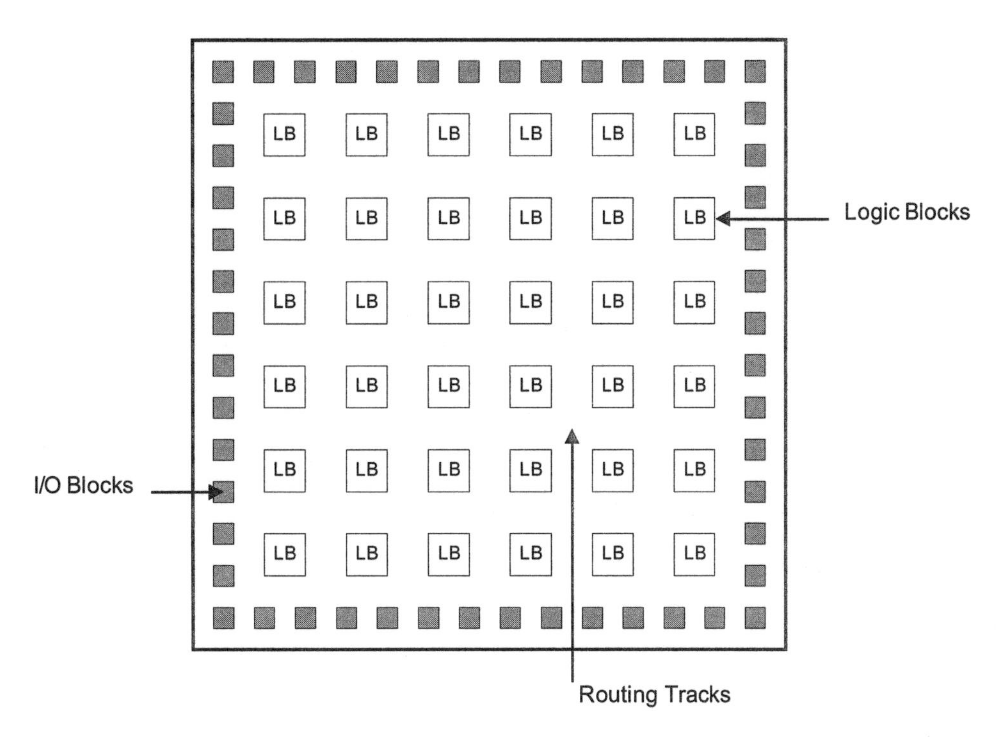

Fig. 4.2 Simplified FPGA Structure

capabilities. Table 4.1 presents examples of where in the PAL-to-gate-array continuum several popular architectures fall.

Programmable-logic-device architectures can be categorized as fine grained or coarse grained. The nature of the structure of the logic and I/O blocks determine whether an architecture is fine or coarse grained.

☞ **Coarse Grained:** Characterized by logic and I/O blocks that contain a relatively large amount of programmable functional-

Table 4.1 PAL-to-Gate Array Continuum

PAL-Like			. . . . **Gate-Array-Like**
Xilinx 9500 CPLD Family	Altera Flex 8000	Xilinx XC3000 and XC4000 Series	Actel Act 1, 2, and 3

ity. These architectures tend to also employ complex routing resources.

☞ **Fine Grained:** Characterized by a core of small, relatively simple logic blocks.

Fine-grained architectures result in devices that begin to resemble gate arrays. Because the structures of fine-grained devices are less complex than those of coarse-grained devices, the precompile, place-and-route analysis of fine-grained devices is more straightforward and the results more predictable than when using coarse-grained devices.

Programmable-logic-device architectures can also be categorized as deterministic versus nondeterministic with respect to the calculation of signal delay prior to place and route. While there is some degree of delay unpredictability prior to place and route in most cases, complex structures and routing resources make delay calculation difficult and sometimes impossible to predict.

The physical realization of the programmable element of a programmable logic device does, to some extent, dictate its architecture. The most common implementations are the antifuse, the static RAM, and the EPROM/EEPROM.

☞ **Antifuse:** Actel uses a proprietary antifuse technology that they call PLICE. These parts exhibit a fine-grained architecture with small logic blocks. Delays through these devices can be predicted with relative accuracy prior to place and route.

☞ **Static RAM:** Many of the products from Xilinx and Altera are based on static RAM technology. The Xilinx devices exhibit a coarse-grained architecture with complex logic blocks, complex routing facilities, and difficult-to-predict delays. Determining routing delays for Xilinx XC3000 and XC4000 series devices is possible only after placement and routing. The Altera devices, on the other hand, are made up of groups of relatively small cells that combine into groups to form larger blocks. Altera's routing resources are structured so that routing delays are predictable prior to place and route.

☞ **Other:** Altera MAX 5000 EPLDs and Xilinx EPLDs both use EPROM cells as their programming technology.

Clock networks may use dedicated or "normal" routed resources. Dedicated clock networks support high performance by providing low (subnanosecond) skew and guaranteed performance. They contain no programming elements in the path from I/O pad driver to block input and are accessed by special I/Os. In some cases clock networks are divided into those that supply clocks to just the core logic blocks, and those that supply clocks to the I/O blocks to further enhance performance. Also, there may be special cells such as RAMs, PLLs, etc., available on the chip. All of these architectural features not only have a direct effect on the performance of the device, but also affect ease of design implementation.

4.2.1 Logic Block

The core of a programmable logic device is made up of an array of logic elements. These are called logic blocks, logic modules, configurable logic blocks, or a variety of other names depending upon the vendor. The complexity of the blocks can range from very simple to relatively complex structures. Generally speaking, logic elements with less complex structures present less complex timing issues. When a device with a simpler architecture does not meet performance goals following placement and routing, the timing-analysis results are more easily interpreted, and corrective actions are more easily implemented.

Logic-block complexity also determines the granularity of a programmable logic device's architecture. This especially applies to FPGAs. A logic block with low-level complexity results in an FPGA with a fine-grained architecture having gate-array-like properties. On the other hand, complex logic blocks result in coarse-gained architectures. One advantage of the small, fine-grained logic-block architecture is that it is a good target for synthesis engines.

Later sections that present the architectures of specific devices provide detailed descriptions of the logic blocks used by these devices.

4.2.2 Input/Output Block

The core of a programmable logic device is surrounded by a set of logic elements that interface between the core and the I/O pads of the device. These are called I/O blocks, I/O elements, or I/O modules. The functionality of these blocks varies less from vendor to vendor than do the logic blocks. Typically they include directional and/or bidirectional buffering, registered inputs and outputs, translation, etc.

4.2.3 Routing Facilities

The interconnection between logic and I/O blocks via the programmable routing facilities varies greatly from vendor to vendor. In some architectures the routing resources are partitioned. The device's array of blocks is subdivided into regions with a set of resources provided to interconnect the blocks within each region. Interconnects between regions are made with another set of resources, or a means is provided to extend the local resources. In either case, a performance penalty is incurred when connections extend beyond a local region.

The complexity of programmable-logic-device timing analysis is due to the difficulty of predicting routing delays. Ideally, from a designer's point of view, it should be possible to determine whether a design's performance goals will be met using a particular device without going through the whole design process. At times it is possible to accurately predict delays in the preliminary stages of a design, but devices from different vendors offer varying degrees of predictability. In some cases an iterative approach is required, using the vendor's development system software to execute trial place-and-route runs.

CPLDs often have routing structures with propagation delays that can be determined prior to placement and routing, but in many cases FPGAs with complex routing resources do not. Elements that affect routability and, to some extent, routing-delay predictability are:

- **Programming Technology:** Various schemes are used to form the interconnect contacts. These include antifuse connects, pass transistors, and active buffers.
- **Complexity:** At the complex end of the spectrum is the Xilinx XC3000/XC4000 family with many different segment-line lengths and switching matrices, and at the other end is Actel with simple antifuse contacts and a simple interconnect scheme.
- **Availability:** A sufficient magnitude of routing resources should be available so that it is possible to utilize a majority of the device's resources while still achieving a reasonable performance goal.

Figure 4.3 is a simplified diagram showing the complex structure of the routing resources used by the Xilinx XC3000/XC4000

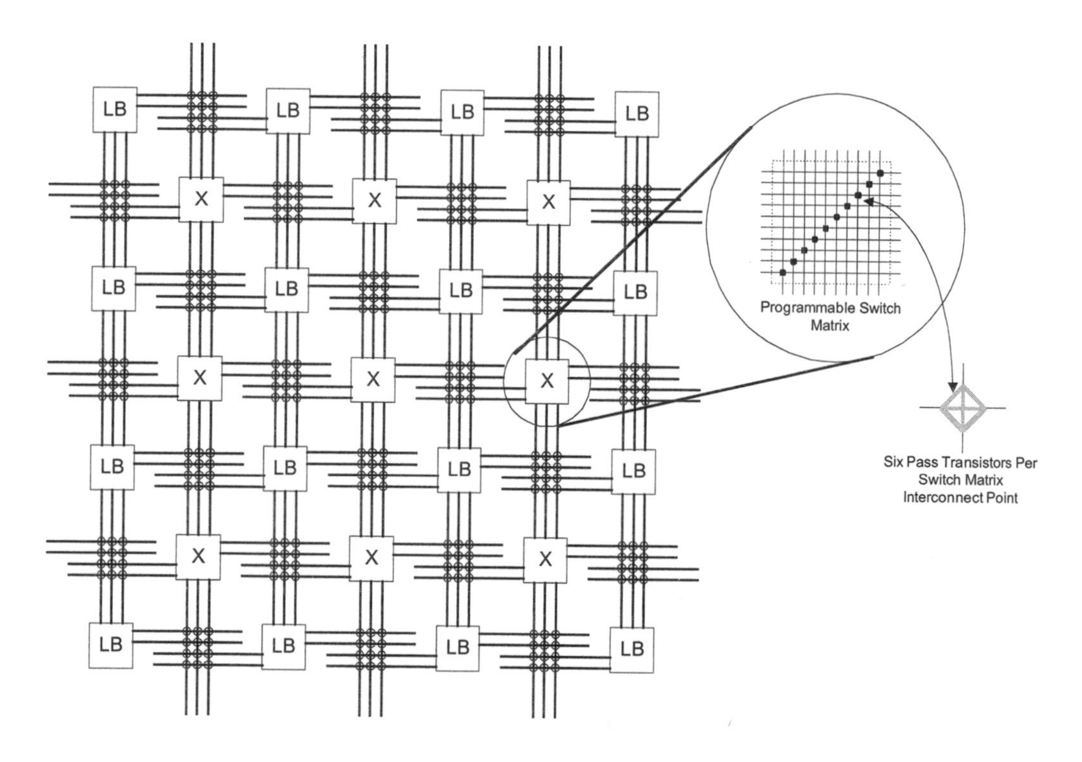

Fig. 4.3 Xilinx X3000/X4000 Family Routing Facilities [Xilinx, 1998] Courtesy of Xilinx, Inc.

families of FPGAs. The less complex scheme used by Actel is shown in Figure 4.4. Actel uses low-resistance antifuse contacts with low-capacitance segmented routing tracks to optimize performance.

4.3 Design Flow

The sequence of steps that makes up a programmable-logic-design flow does not vary much from vendor to vendor. These steps are shown in Figure 4.5.

Design entry can be done using FPGA vendor-specific tools, but is generally done using third-party tools. At this stage of develop-

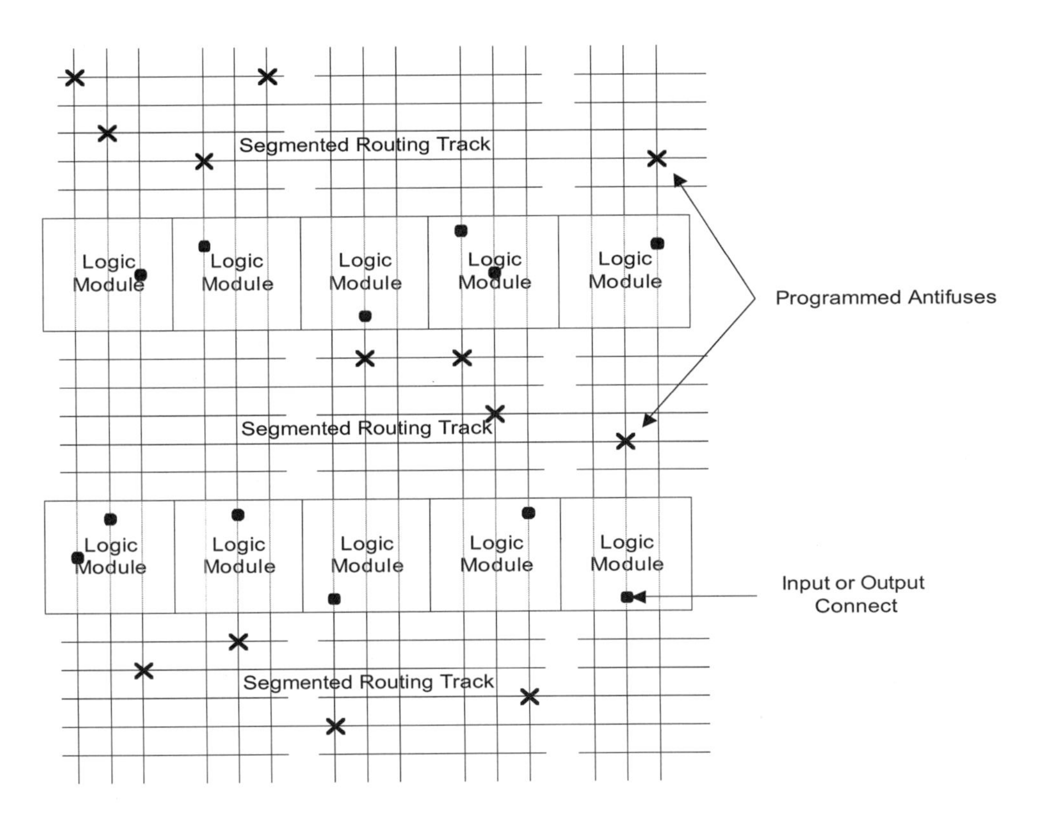

Fig. 4.4 Actel ACT Family Routing Facilities [Actel, 1996] Courtesy of Actel Corporation

ment, performance estimates may be made using vendor data sheets and timing models. Preliminary device selection and design trade-offs are made based on the estimated performance information.

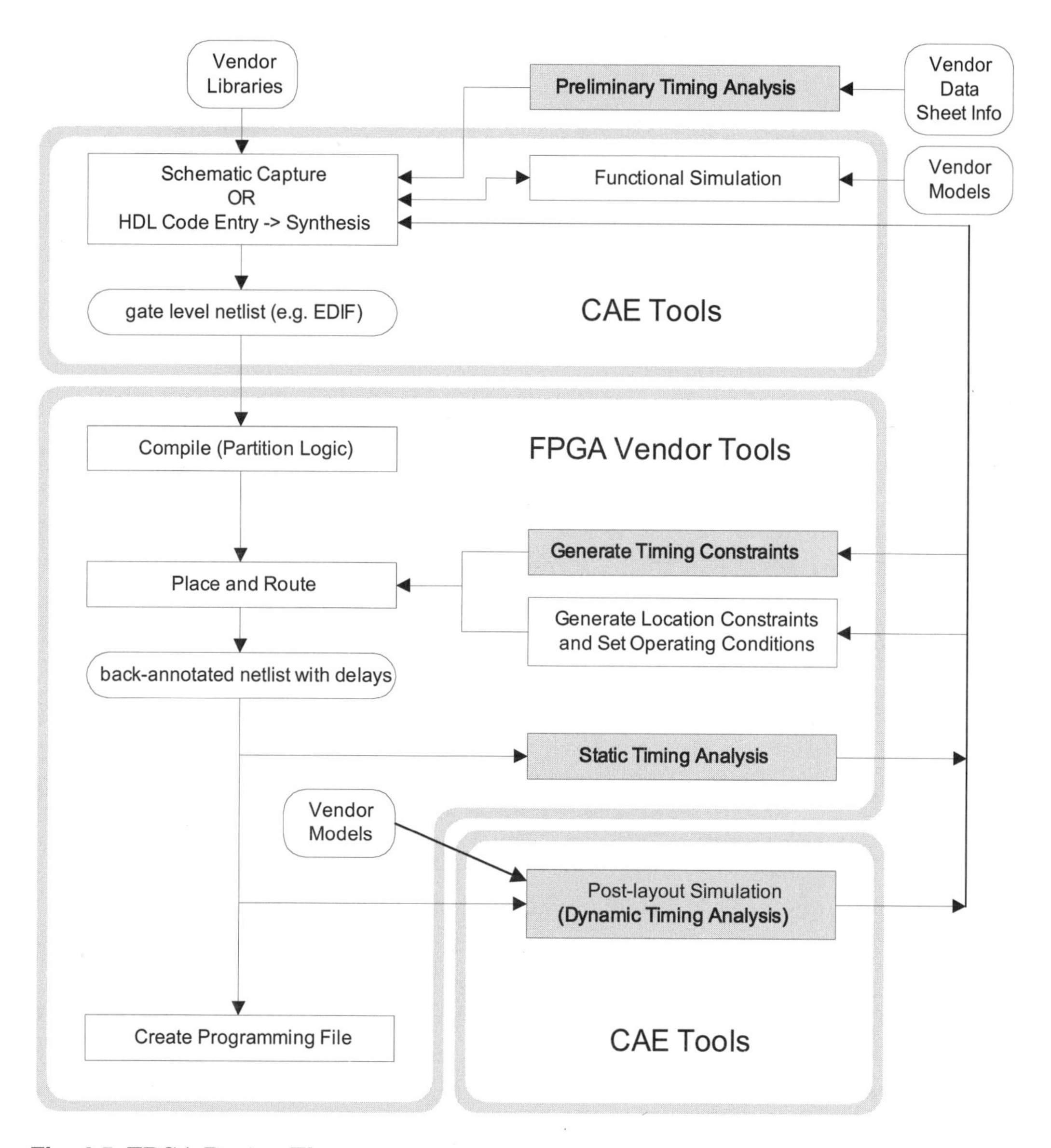

Fig. 4.5 FPGA Design Flow

Once the design is developed at the functional level, either through schematic capture or HDL synthesis, netlists are generated. For example, an EDIF netlist may serve as the database for the vendor's place-and-route software while a verilog or VHDL netlist is required by the simulation engine. Simulation verifies that the design logically functions properly at the gate level. This is especially important if the design was captured using HDL synthesis. Once the routing is done and the route delays are fed back into the gate-level model, the simulation can determine if the design meets the timing goal. However, most design flows depend on static timing analysis for final timing verification. The netlist may need translating from the format used by the CAE tools into the format required by the vendor place-and-route tools. EDIF is a format commonly used by vendors.

The vendor-specific tools (vendor development system) are used to create the physical implementation of the design. The place-and-route process is guided by constraints placed on the design. These constraints can take the form of location constraints and/or timing constraints. Constraints are made available to the place-and-route software via a file that the designer creates, or constraints can be entered interactively using the development system's graphical user interface. Also, constraints can be conveyed to the place-and-route software via attributes placed on schematics or in HDL code. Some typical timing constraints are the system clock period, input pin arrival times, and output pin required times. In addition to constraints there may be various user-configurable parameters that control place-and-route strategies.

It is best to start by not constraining the design, but telling the tool to optimize in favor of area rather than timing. The results are evaluated and if the design does not meet expectations, constraints will need to be applied and strategy parameters set in such a way that the tool works to improve the results. If a point is reached where the results can no longer be improved, and the desired area-performance goals have not been attained, two possibilities remain. Redesign at the functional level may be required. One example would be to reduce the number of logic levels between registers. The other possibility is to use a higher-performance device in the same family, or perhaps target a device with architecture more suited to the design under development.

Following placement and routing, the vendor's development system can generate a standard delay format (SDF) file, which contains the delay of every path. The SDF file can be used to back annotate the delays into the simulation file or final timing analysis can be done with static timing analysis (STA). Timing verification using the simulator is not as robust as STA because it can only verify that the design works properly for those paths that are exercised. Static timing analysis can verify that all paths in the design meet the performance goals.

4.4 TIMING PARAMETERS

Timing parameters define the delays inherent in each component that makes up a programmable logic device. They include numbers published by vendors as part of device data sheets and are valid over specified operating conditions. Timing parameters can be categorized as internal or external. The timing delays contributed by individual architectural elements are internal timing parameters. For example, the delay through an internal combinatorial block of logic is an internal timing parameter. These timing parameters cannot be measured explicitly. External timing parameters represent actual pin-to-pin timing characteristics. Each external timing parameter consists of a combination of internal delay elements. "They are worst-case values, derived from performance measurements and are ensured by device testing or characterization" [Altera98].

As Actel notes in their "FPGA Data Book and Design Guide," programmable-logic-device timing parameters can be categorized as family dependent, device dependent, or design dependent. For example, I/O buffer characteristics may be common to all members of a particular family. On the other hand, internal routing delays are device dependent. Design dependency means actual routing delays are not determined until after placement and routing of the design is complete.

Design dependency is the most difficult thing to deal with when making programmable-logic-device performance estimates. FPGA architectures that use complex routing schemes fall into the category of unpredictability. On the other hand, CPLDs usually have predictable routing structures. For example, the Xilinx XC9500 family uses a predictable-delay switching-matrix scheme that fully interconnects the logic blocks of the device.

4.4.1 Timing Derating Factors

The majority of programmable logic devices are implemented using a CMOS process, which means their performance varies according to voltage, temperature, and process. When performing timing analysis, the designer is usually concerned with worst-case performance. The designer needs to know if the timing parameters specified for a programmable logic device permit a design to be implemented that meets the required performance goals. Additionally, the designer may need to be aware of best-case performance, especially in the case of outputs where hold time becomes an issue with regard to elements being driven by the device. Table 4.2 shows how voltage, temperature, and process affect performance.

Vendors supply data for devices operating under a defined set of operating conditions, e.g., Actel uses worst-case commercial condition. (See Table 4.3 for definitions of operating conditions.) Variance from the vendor-specified conditions is accounted for by derating factors. Derating factors for Actel ACT 3 devices are shown in Table 4.4. Actel's base case uses a junction temperature of 70°C

Table 4.2 Performance for CMOS Devices as a Function of Voltage, Temperature, and Process

Performance	Voltage	Temperature	Process
Best-Case	Maximum	Minimum	Fast
Worst-Case	Minimum	Maximum	Slow

Table 4.3 Actel ACT 3 Temperature and Voltage Derating Factors (Normalized to Worst-Case Commercial Conditions, TJ = 4.75 V, 70°C) [Actel, 1996] Courtesy of Actel Corporation. Portions reprinted with permission.

	Temperature TJ (junction/°C)						
VDD (V)	**–55**	**–40**	**0**	**25**	**70**	**85**	**125**
4.50	0.72	0.76	0.85	0.90	1.04	1.07	1.17
4.75	0.70	0.73	0.82	0.87	**1.00**	1.03	1.12
5.00	0.68	0.71	0.79	0.84	0.97	1.00	1.09
5.25	0.66	0.69	0.77	0.82	0.94	0.97	1.06
5.50	0.63	0.66	0.74	0.79	0.90	0.93	1.01

Table 4.4 Actel ACT 3 Operating Condition Derating Factors (Normalized to Worst-Case Commercial Conditions) Courtesy of Actel Corporation. Portions reprinted with permission.

Operating Condition	Worst-Case		Best-Case	
	Definition	**Derating Factor**	**Definition**	**Derating Factor**
Commercial	4.75V, 70°C	**1.00**	5.25V, 0°C	Not listed
Industrial	4.50V, 85°C	0.66	5.50V, –40°C	1.07
Military	4.50V, 125°C	0.63	5.50V, –55°C	1.17

and a supply voltage (VDD) of 4.74V. If the designer knows that the device will not be used in such pessimistic conditions, he may choose to simulate at a lower junction temperature and a higher voltage. With a junction temperature of 0°C and a supply voltage of 5.0V, the derating factor is 0.79. All delays are affected by the derating factor, which means the device operates faster at the new operating point because the derating factor is less than 1.0.

While temperature and voltage derating factors are often available, process variation numbers are difficult to obtain. In an application note, Xilinx indicates that the difference between their slowest and fastest process is 40% [Smith, 1997].

4.4.2 Grading Programmable Logic Devices by Speed

Programmable-logic-device vendors sort their parts according to speed. For example, Actel measures performance with a special binning circuit that consists of an input buffer driving a string of buffers or inverters followed by an output buffer [Smith97]. The parts are sorted from measurements on the binning circuit according to logic module propagation delay. The propagation delay, t_{PD}, is defined as the average of the rising (t_{PLH}) and falling (t_{PHL}) propagation delays of a logic module

$$t_{PD} = \frac{(t_{PLH} + t_{PHL})}{2} \qquad \textbf{(Eq. 4.1)}$$

Table 4.5 shows how performance varies with speed grade for Actel ACT 3 devices. The quantities in parentheses show delay relative to the delay at speed grade Std. Note that module propagation delay and routing delays are affected similarly.

There is no standard method of designating speed grades in programmable-logic parts. Some examples are listed in Table 4.6.

The speed-grade code Xilinx uses for their XC9500 CPLD family is tied to the logic block t_{PD} timing parameter; e.g., a speed grade of 7 corresponds to t_{PD} = ~7.0ns. Also note that programmable logic

Table 4.5 ACT 3 Logic Module and Routing Delays as a Function of Speed Grade and Fanout (Worst-Case Commercial Conditions) [Actel Corporation, 1996] Courtesy of Actel Corporation. Portions reprinted with permission.

		Predicted Routing Delays				
Speed Grade	**Logic Module Prop Delay (t_{PD})**	**Fanout = 1 (t_{RD1})**	**Fanout = 2 (t_{RD2})**	**Fanout = 3 (t_{RD3})**	**Fanout = 4 (t_{RD4})**	**Fanout = 8 (t_{RD8})**
3	2.0 (0.67)	0.9 (0.70)	1.2 (0.67)	1.4 (0.67)	1.7 (0.68)	2.8 (0.67)
2	2.3 (0.77)	1.0 (0.77)	1.4 (0.78)	1.6 (0.76)	1.9 (0.76)	3.2 (0.76)
1	2.6 (0.87)	1.1 (0.85)	1.6 (0.89)	1.8 (0.86)	2.2 (0.88)	3.6 (0.86)
Std	3.0 (1.00)	1.3 (1.00)	1.8 (1.00)	2.1 (1.00)	2.5 (1.00)	4.2 (1.00)

Table 4.6 Speed Grade Designations

Relative Speed	Actel ACT Families	Altera FLEX 8000	Xilinx XC9500 CPLD	Xilinx XC3000/ XC4000
Slower	Std	A-4	20	–4
·	1	A-3	15	–3
·	2	A-2	10	–2
·	3		7	–1
Faster			6	–09

devices are sometimes "down-binned." This practice results in a faster part being shipped against an order for a slower part.

4.4.3 Best-Case Delay Values

Generally all of the timing parameters reported by vendors are some variation on worst-case delay values. Best-case values are difficult, if not impossible, to obtain. This is because they are not easy to test for, plus every time the fabrication process for a device is changed its best-case delay values change. Best-case delay values are used to determine flip-flop hold-time violations. Hold-time violations between flip-flops on the same chip are avoided by careful design of the on-chip clock distribution network. When the worst-case clock-skew value is shorter than the sum of minimum clock-to-Q delay plus the minimum interconnect delay, there will not be a hold-time violation. The vendor development software should guarantee that no flip-flop hold-time violations are created when a design is placed and routed.

It is, however, up to the designer to prevent chip-to-chip hold-time problems. If the receiving device has a hold-time requirement, the problem is exasperated. Some FPGA vendors (for example, Xilinx) have attempted to mitigate this problem by ensuring that there is not a positive input clock to input data hold-time requirement. They have done this by adding delay to the input data paths.

In an application note Xilinx sheds some light on best-case delay calculation [Xilinx99]. The calculated figures that are given are:

- ☞ Tester guardband: 10%
- ☞ Voltage (4.75V is worst case -> 5.25V is best case): 10%
- ☞ Temperature (85°C is worst case -> 0°C is best case): 30%
- ☞ Process: 40%

Therefore:

best-case delay = 0.9 * 0.9 * 0.7 * 06 * worst-case delay

best-case delay = 0.34 * worst-case delay

It is further suggested that to be very conservative one should use the following:

best-case delay = 0.25 * (worst-case delay for fastest available speed grade)

4.5 Timing Analysis

When the architecture of a programmable logic device allows predictable delay calculations to be made prior to completing place and route, the vendor may provide a timing model for that family. A timing model provides the timing characteristics for the various paths within and through a device. The characteristics are indicated as timing parameters on the model, and correspond to values listed on the data sheets for individual members of a family.

A preliminary timing analysis can be performed prior to placement and routing when one understands the architecture of a device and the vendor provides a timing model with appropriate timing parameters. The following sections give detailed descriptions for several types of vendor-specific programmable logic devices. Architectural descriptions are given along with timing models and sample timing calculations.

The necessity of postplace-and-route timing verification is not obviated by preliminary timing analysis. It is still of utmost importance to verify the design using the timing-analysis tools provided by the vendor's software development system. Vendors typically indicate in their timing model or timing parameter information that results are for typical cases and that accurate results can only be achieved via their development system.

4.5.1 Actel ACT FPGA Family

The structure of programmable logic devices in the Actel ACT FPGA family is similar to that of a conventional gate array. The core consists of an array of logic modules used to implement combinatorial and sequential logic. The logic modules are of low complexity, and are interconnected with segmented routing tracks. The segment lengths are predefined and are connected with low-impedance antifuse elements to create the particular routing length required of the interconnect signal (see Figure 4.4). Surrounding the logic core are I/O modules that translate and interconnect the logic signals from the core to the I/O pads.

4.5.2 Actel ACT 3 Architecture

The basic architectural elements of the Actel ACT 3 family of FPGA devices are shown in Figure 4.6. The fine-grained architecture uses antifuse interconnects and simple logic blocks. Rows consisting of two kinds of logic blocks alternate with routing channels. I/O modules form the periphery of the array.

The two types of logic blocks are shown in detail. The C-module is designed to implement combinatorial macros. Using a four-to-one multiplexer with additional logic at the multiplexer's select inputs allows a variety of combinatorial functions to be implemented. The S-module is designed to implement high-speed sequential functionality. Each S-module consists of the set of C-module combinatorial logic elements driving a flip-flop.

For more detail on the Actel ACT 3 family of FPGA devices, including descriptions of the I/O block, clock networks, and routing resources, see Actel Corporation, 1996, in References.

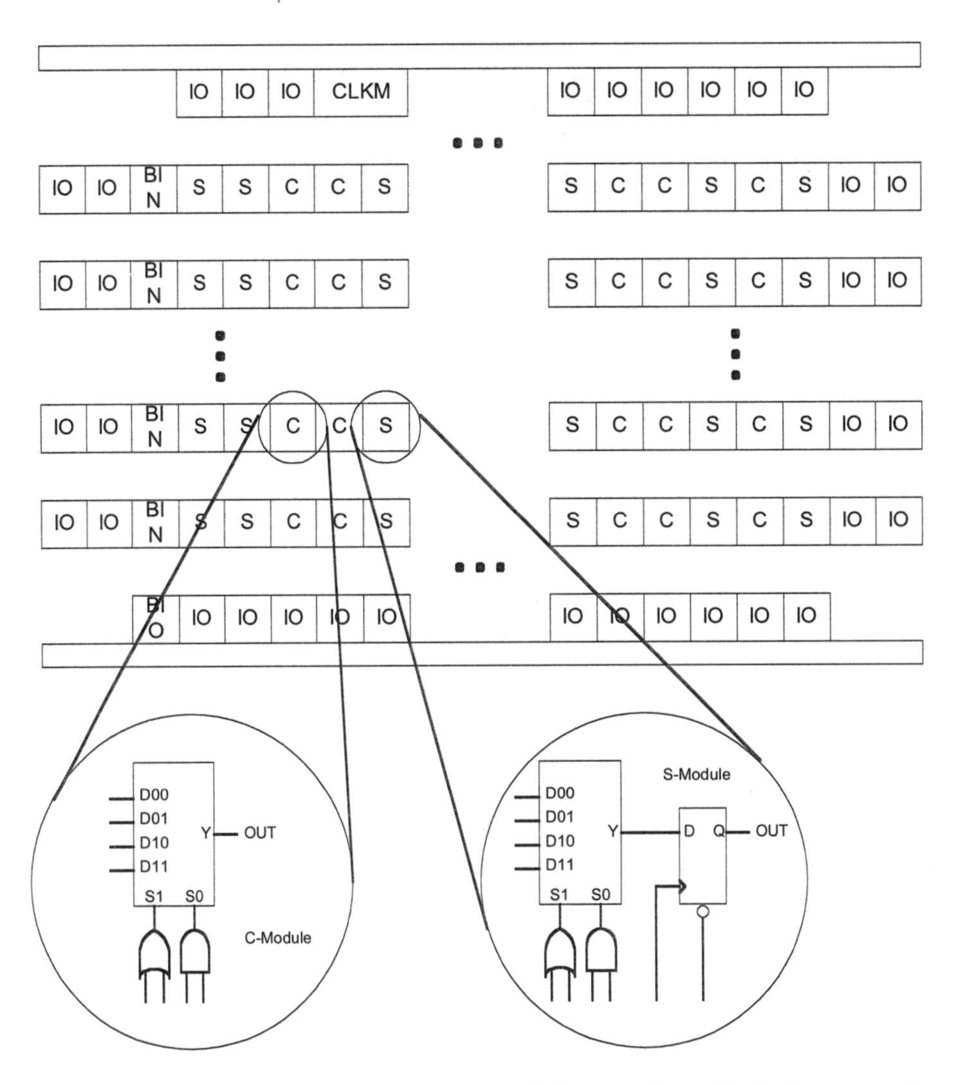

Fig. 4.6 Actel ACT 3 Array Structure [Actel Corporation, 1996] Courtesy of Actel Corporation

4.5.3 Actel ACT 3 Timing Model

The timing model for an Actel ACT 3 part is shown in Figure 4.7. The delay of all basic elements (I/O module, logic module, routing) is given including load-dependent routing delay. Routing delay information is included in the t_{RDn} and t_{IRDn} timing parameters. These

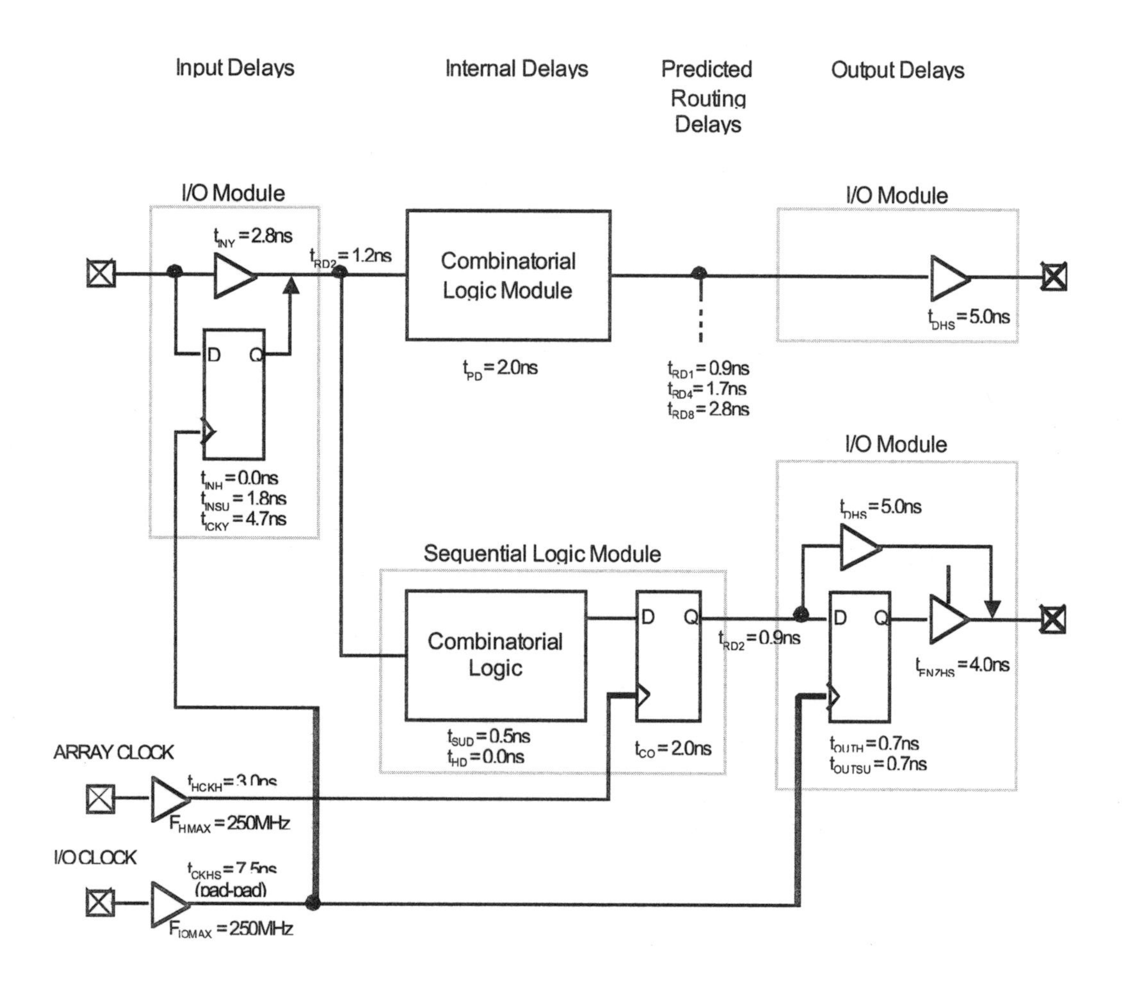

Fig. 4.7 Actel ACT 3 Timing Model (Values Shown for Actel A1425A-3 at Worst Case Commercial Conditions) [Actel Corporation, 1996] Courtesy of Actel Corporation

delays depend on the resistive/capacitive loading of the routing tracks, the antifuse connect elements, and the number of module inputs being driven. The delay increases as the length of routing tracks, the number of interconnect elements, or the number of inputs increases. The routing track length and number of interconnect elements can be statistically correlated to the number of input loads driven by an output.

Routing delays can be predicted because Actel's interconnect scheme minimizes the number of antifuse connects (e.g., 90% of the paths use 2), and the types (i.e., lengths) of routing tracks are also kept to a minimum. However, when long tracks are used, their delay can vary from 4ns to 14ns. Long tracks are routing resources that span multiple rows, columns, or modules and can use three or four antifuse connections.

Table 4.7 Actel ACT Family Timing Parameter Definitions [Actel Corporation, 1996] Courtesy of Actel Corporation. Portions reprinted with permission.

Timing Parameter	Timing Parameter Definition
F_{MAX}	Flip-flop clock frequency
t_{HCKH}	Dedicated array clock network: Input low to high (pad to S-Module input)
t_{CKHS}	IOCLK pad to pad H/L hi slew
t_{INH}	Input buffer latch hold
t_{INSU}	Input buffer latch setup
t_{ICKY}	Input register IOCLK pad to Y
t_{IRDn}	Input module predicted routing delay for fanout of n
t_{PD}	Module propagation delay
t_{SUD}	Flip-flop data input setup
t_{HD}	Flip-flop data input hold
t_{CO}	Sequential clock to Q
t_{RDn}	Predicted routing delay for fanout of n
t_{DLS}	Data to pad, low slew
t_{OUTH}	Output buffer latch hold
t_{OUTSU}	Output buffer latch setup

Example 4.1

Figure 4.8 shows an internal path through an Actel ACT 3 A1425A device along with the relevant timing parameters. Values given are for worst-case commercial operating conditions using a part with a speed grade of "Std." The worst-case delay is then,

Example 4.1 (Continued)

$t_{PATH\_WCC} = t_{CO} + t_{RD4} + t_{PD} + t_{RD1} + t_{SUD}$
$= 3.0ns + 2.5ns + 3.0ns + 1.3ns + 0.8ns$
$= 10.6ns$

To compute the delay under other conditions the derating factors of Table 4.5 or 4.6 are used. For example, the delay for best-case industrial (BCI) is calculated as follows:

$t_{PATH\_BCI} = t_{PATH\_WCC}$ * BCI Derating Factor * Process Derating Factor.
$t_{PATH\_BCI} = 10.6 * 0.66 * 0.60 = 4.2ns$

The Actel data book notes that the published routing delays are for typical designs, that they should be used only for estimating device performance, and that postroute timing analysis or simulation is required to determine actual worst-case performance.

Another approach that can be taken to estimate the performance of Actel devices is outlined in an application note, "Estimating Performance and Capacity of Actel Devices" [Actel96]. This approach uses timing estimates for common functional blocks such as counters, state machines, etc.

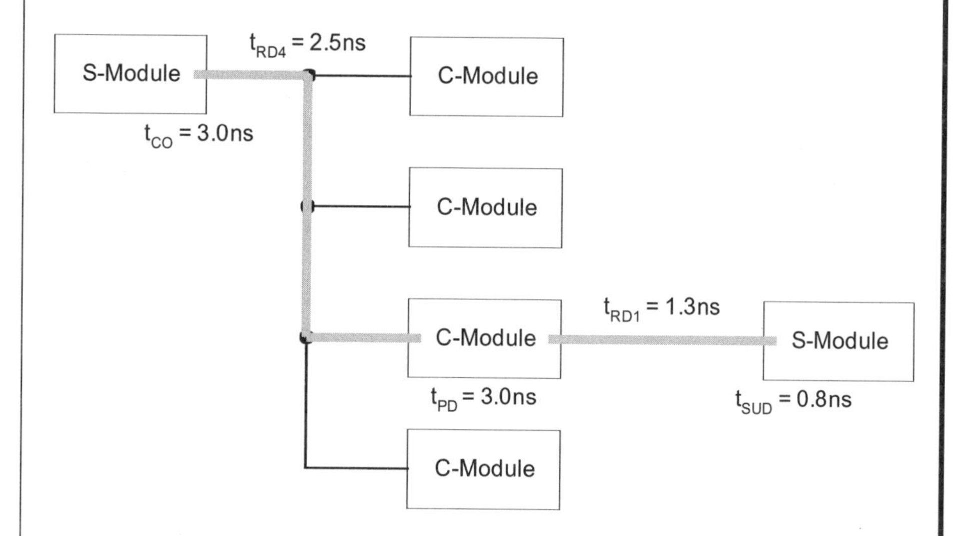

Fig. 4.8 Actel ACT 3 Example Timing Calculation

4.5.4 Altera FLEX 8000

Altera characterizes their FLEX 8000 family of programmable logic devices as "offering fast, predictable interconnect delays," and combining "the high-speed predictable timing, and ease-of-use of EPLDs with the high register count, low standby power, and in-circuit reconfigurability of FPGAs" [Altera98]. The devices in this family range in size from 2,500 to 16,000 gates.

4.5.5 Altera FLEX 8000 Architecture

The logic core of the FLEX 8000 employs a two-level hierarchical structure. At the lower level is an array of fine-grained logic blocks called logic elements (LEs). The logic elements are formed into groups of eight creating a second-level hierarchical structure called the logic array block (LAB). Each LAB is an independent structure with common inputs, interconnections, and control signals. The LAB architecture provides a coarse-grained structure for high device performance and easy routing.

A diagram of the FLEX 8000 architecture is shown in Figure 4.9. The LABs are arranged into rows and columns, with the I/O pins supported by I/O elements (IOEs) located at the ends of these rows and columns. Each IOE contains a bidirectional I/O buffer and a flip-flop that can be used as either an input or output register. LAB-to-LAB and LAB-to-IOE interconnects are provided by the FastTrack interconnects that run the full length and width of the device.

Each LAB consists of eight LEs, their associated carry and cascade chains, LAB control signals, and the LAB local interconnect. Each LE contains a four-input look-up table (LUT), a programmable flip-flop, a carry chain, and a cascade chain. The flip-flop can be configured for D, T, JK, or SR operation, and can be bypassed for combinatorial functions. The cascade chain allows the inputs from several logic elements to be grouped together for high fan-in logic functions with a delay of less than 1ns per stage. The carry chain allows implementation of fast counter and adder functions.

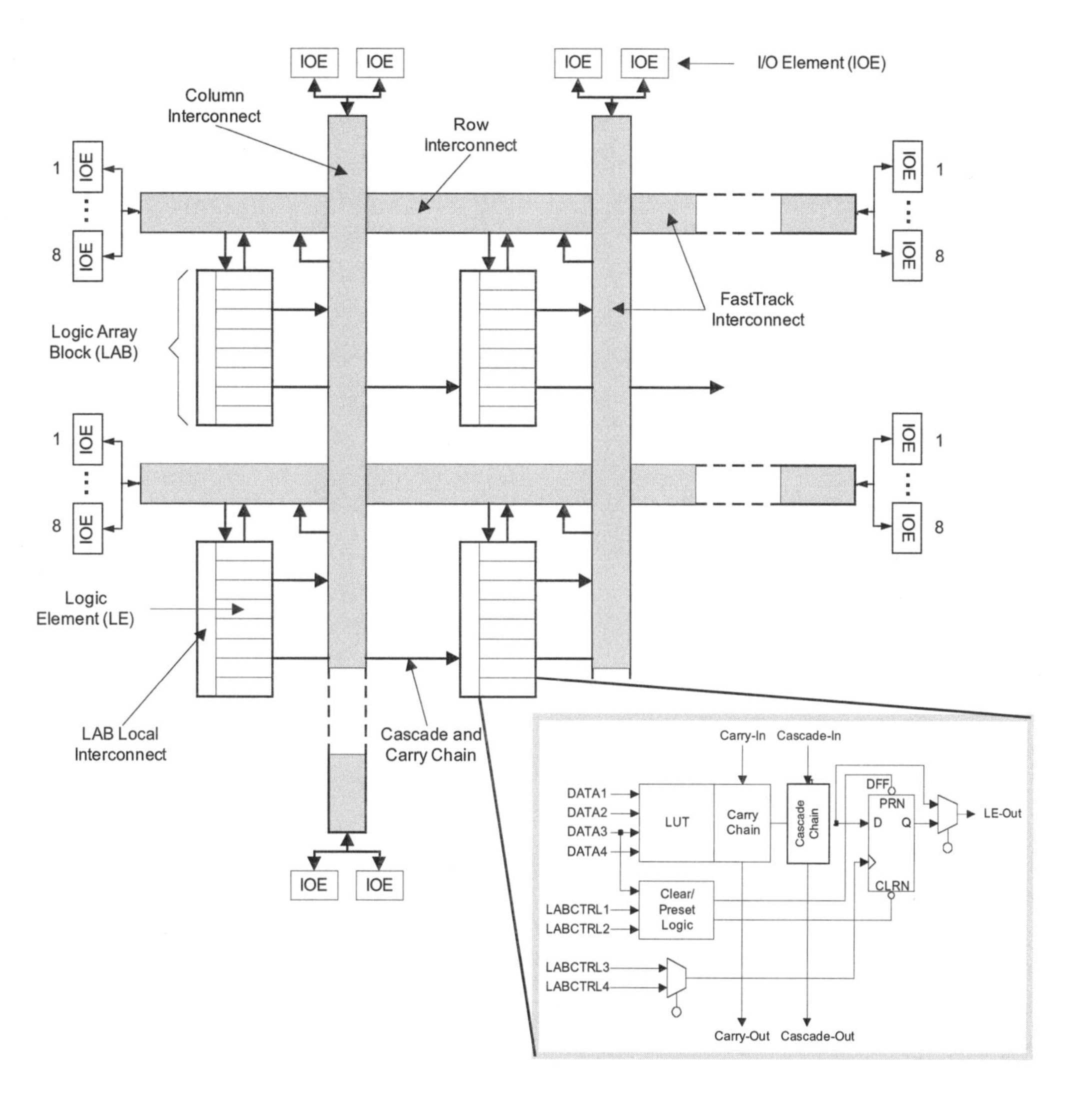

Fig. 4.9 Altera FLEX 8000 Array Structure [Altera Corporation, Sept. 1998] Courtesy of Altera Corporation.

4.5.6 Altera FLEX 8000 Timing Model

The performance of the Altera FLEX 8000 family is predictable prior to place and route. The timing model for the Altera FLEX 8000 family is shown in Figure 4.10 with the various timing parameters defined in Table 4.8. LE, IOE, and interconnect timing parameters are shown.

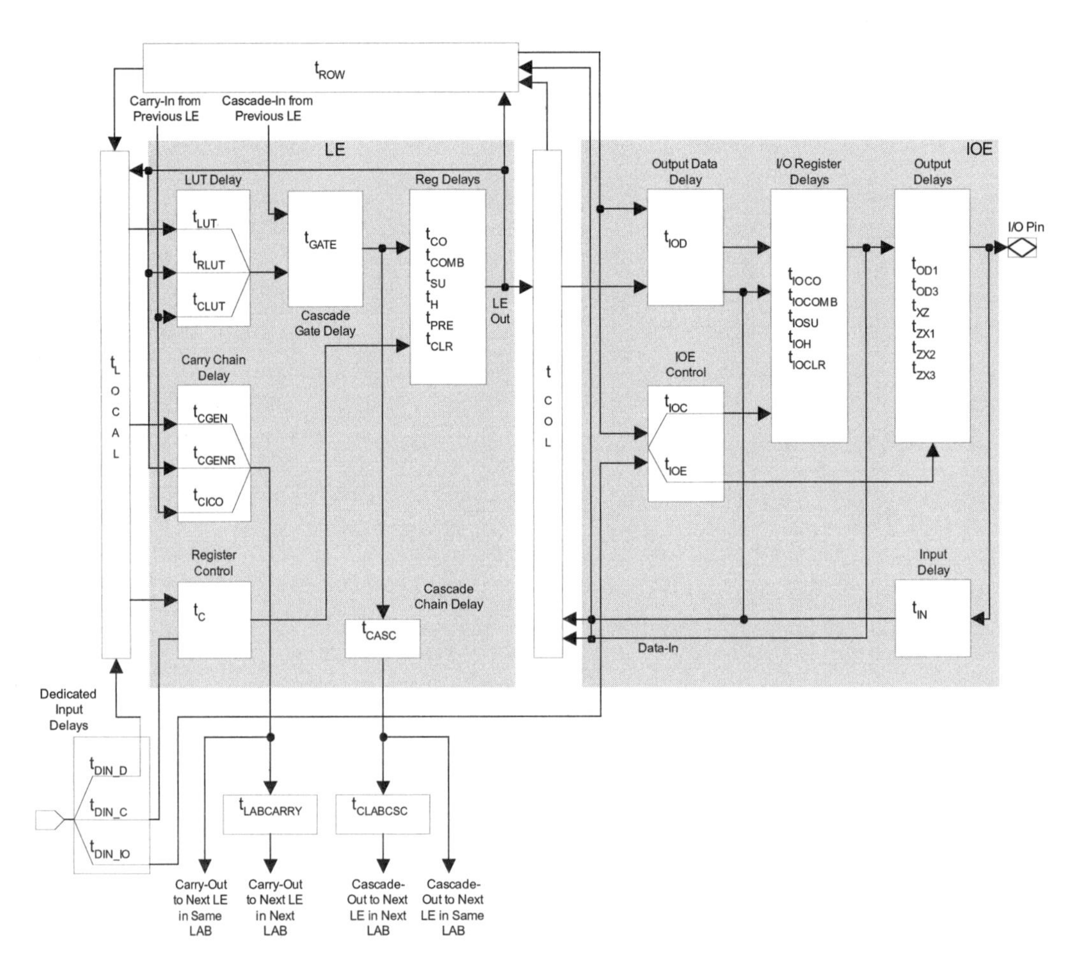

Fig. 4.10 Altera FLEX 8000 Timing Model [Altera Corporation, Jan. 1998] Courtesy of Altera Corporation. Copyright © Altera Corporation 1998. All rights reserved.

Table 4.8 Altera FLEX 8000 Timing Parameters [Altera Corporation, Sept. 1998, and Altera98] Courtesy of Altera Corporation. Copyright © Altera Corporation 1998. All rights reserved.

Timing Parameters		EPF81500A A-3 Speed Grade Min (ns)	EPF81500A A-3 Speed Grade Max (ns)
I/O Element Timing Parameters			
t_{IN}	I/O input pad and buffer delay		1.6
t_{IOD}	Output data delay		0.8

Table 4.8 Altera FLEX 8000 Timing Parameters [Altera Corporation, Sept. 1998, and Altera98] Courtesy of Altera Corporation. (Continued)

Timing Parameters		EPF81500A A-3 Speed Grade	
		Min (ns)	Max (ns)
t_{IOC}	IOE control delay		1.8
t_{IOE}	IOE output enable delay		1.8
t_{IOCO}	I/O register clock-to-output delay		1.0
t_{IOCOMB}	I/O register bypass delay		0.2
t_{IOSU}	I/O register setup time	1.6	
t_{IOH}	I/O register hold time	0.0	
t_{IOCLR}	I/O register clear delay		1.2
t_{OD1}	Output buffer and pad delay with slow slew-rate logic option turned off and V_{CCIO} = 5.0V		1.4
t_{OD2}	Output buffer and pad delay with slow slew-rate logic option turned off and V_{CCIO} = 3.3V		1.9
t_{OD3}	Output buffer and pad delay with slow slew-rate logic option turned on and V_{CCIO} = 5.0V or 3.3V		4.9
t_{XZ}	Output buffer disable delay		1.6
t_{ZX1}	Output buffer enable delay with slow slew-rate logic option turned off and V_{CCIO} = 5.0V		1.6
t_{ZX2}	Output buffer enable delay with slow slew-rate logic option turned off and V_{CCIO} = 3.3V		2.1
t_{ZX3}	Output buffer enable delay with slow slew-rate logic option turned off and V_{CCIO} = 5.0V or 3.3V		5.1
Interconnect Timing Parameters			
t_{DIN_D}	Dedicated input data delay		8.2
t_{DIN_C}	Dedicated input control delay		5.0
t_{DIN_IO}	Dedicated input I/O control delay		5.0
t_{COL}	FastTrack interconnect column delay		3.0
t_{ROW}	FastTrack interconnect row delay		6.2

Table 4.8 Altera FLEX 8000 Timing Parameters [Altera Corporation, Sept. 1998, and Altera98] Courtesy of Altera Corporation. (Continued)

Timing Parameters		EPF81500A A-3 Speed Grade	
		Min (ns)	Max (ns)
t_{LOCAL}	Local interconnect delay		0.5
$t_{LABCARRY}$	Carry chain delay to the next LAB		0.3
$t_{LABCASC}$	Cascade chain delay to the next LAB		0.3
Logic Element Timing Parameters			
t_{LUT}	LUT delay		2.5
t_{RLUT}	LUT using LE feedback delay		1.1
t_{CLUT}	Carry chain LUT delay		0.0
t_{CGEN}	Carry-out generation delay		0.5
t_{CGENR}	Carry-out generation using LE feedback delay		1.1
t_{CICO}	Carry-in to carry-out delay		0.5
t_{C}	Register control delay		2.0
t_{GATE}	Cascade gate delay		0.0
t_{CASC}	Cascade chain delay		0.7
t_{CO}	LE clock-to-output delay		0.5
t_{COMB}	Combinatorial output delay		0.5
t_{SU}	LE register setup time	1.1	
t_{H}	LE register hold time	1.1	
t_{PRE}	LE register preset delay		0.7
t_{CLR}	LE register clear delay		0.7
t_{CH}	Minimum LE register clock-high time	4.0	
t_{CL}	Minimum LE register clock-low time	4.0	
External Timing Parameters			
t_{DRR}	Register-to-register delay		20.1
t_{ODH}	Output data hold time after clock	1.0	

Example 4.2

Figure 4.11 illustrates two paths through an Altera FLEX 8000 device using simplified versions of Figure 4.10, showing only those elements of the timing model that pertain to these paths. The first path shown in Figure 4.11(a) extends from a row input pin, through combinatorial logic, and ends at a column output pin. The delay calculation for this path is,

$$t_{COMBPATH} = t_{IN} + t_{ROW} + t_{LOCAL} + t_{LUT} + t_{GATE} + t_{COMB} + t_{COL} + t_{IOD} + t_{IOCOM} + t_{OD1}$$

In the second path, Figure 4.11(b), data is entering the device at a column input pin and is clocked into a register by a global clock. The data setup time calculation is,

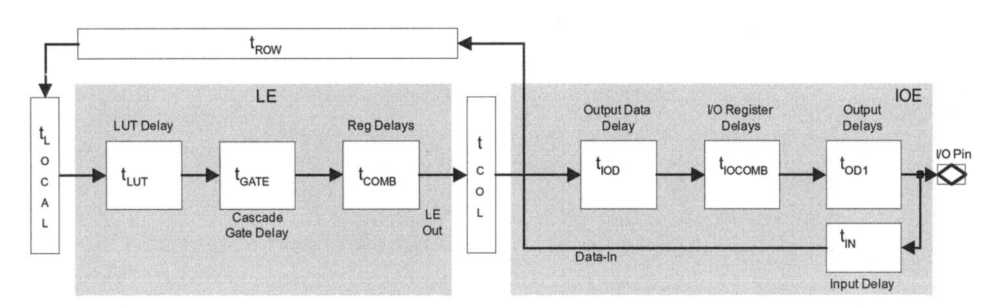

a) Combinatorial Path

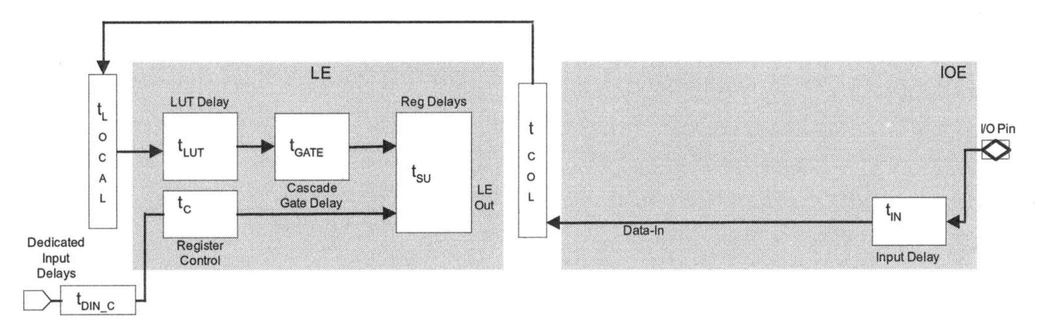

b) Data in to Register Setup Path

Fig. 4.11 Altera FLEX 8000 Timing Model, Simplified for Example 4.2

Example 4.2 (Continued)

$t_{DSU} = (t_{IN} + t_{COL} + t_{LOCAL} + t_{LUT} + t_{GATE}) - (t_{DIN\_C} + t_C) + t_{SU}$

For the EPF81500A A-3 part this gives,

t_{COMB} = 1.6ns + 6.2ns + 0.5ns + 2.5ns + 0.0ns + 0.5ns + 3.0ns + 0.8ns + 0.2ns + 1.4ns = 16.7
t_{DSU} = (1.6ns + 3.0ns + 0.5ns + 2.5ns + 0.0ns) – (5.0 + 2.0) + 1.1 = 1.7ns.

In the second case, notice that the delay in the clock network offsets much of the delay in the data path to the register input and reduces input data setup-time requirements. The data sheet gives only maximum values for t_{DIN_C} and t_C and these are used in the above calculation. But minimum values should be used for these parameters in this case; i.e., the sooner the clock gets to the register the less time data has to propagate to the register's input.

4.5.7 Xilinx XC3000 /XC4000 FPGA Families

The XC3000/XC4000 FPGA family of devices are, as described by Xilinx in *The Programmable Logic Data Book* [1998]: "Implemented with a regular, flexible, programmable architecture of configurable logic blocks (CLBs), interconnected by a powerful hierarchy of versatile routing resources, and surrounded by a perimeter of programmable input/output blocks (IOBs). They have generous routing resources to accommodate the most complex interconnect patterns."

Because there is a great deal of flexibility and because the routing resources are quite complex (see Figure 4.3), estimating performance is difficult. Xilinx does not provide a simple timing model for devices in these families. CLB-switching-characteristic guidelines, IOB input- and output-switching-characteristic guidelines, and pin-to-pin input- and output-parameter guidelines are provided to give a rough idea of the performance capabilities of a particular device.

To begin to understand the complexity of the Xilinx devices, consider the "simplified" block diagram of a Xilinx configurable logic block shown in Figure 4.12. A portion of the XC4000E CLB switching characteristic guidelines are shown in Table 4.9. They are called guidelines because, as Xilinx indicates, for "specific, more precise, and worst-case guaranteed data" one must use the values reported by the static timing analyzer or the back-annotated simulation netlist [Xilinx, 1998].

Another problem with trying to calculate the path delays in these parts is that the routing delays are not deterministic until a design has been placed and routed. The delays within CLBs and

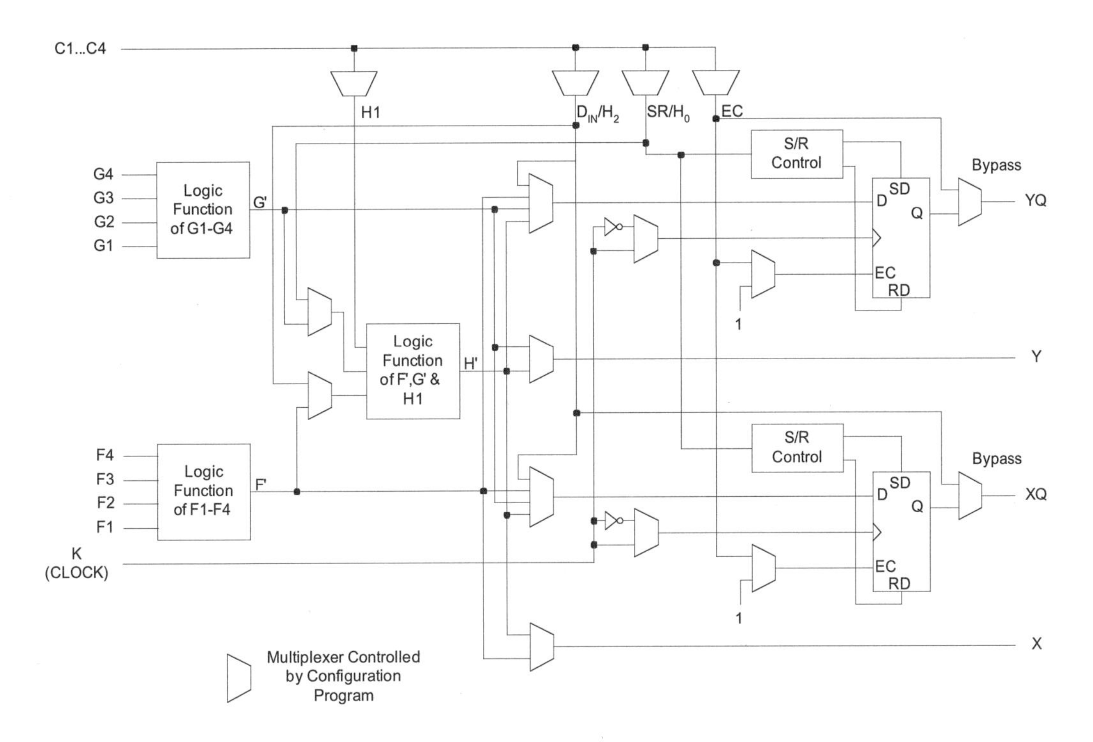

Fig. 4.12 Simplified Block Diagram of the Xilinx XC4000 Series CLB [Xilinx, 1998] Courtesy of Xilinx, Inc.

Table 4.9 Xilinx XC4000E CLB Switching Characteristic Guidelines (partial listing) [Xilinx, 1998] Courtesy of Xilinx, Inc.

Speed Grade		**–3**		
Description	**Symbol**	**Min**	**Max**	**Units**
Setup Time Before Clock K				
F/G inputs	T_{ICK}	3.0		ns
F/G inputs via H	T_{INHCK}	4.6		ns
C inputs via H0 through H	T_{HH0CK}	3.6		ns
C inputs via H1 through H	T_{HH1CK}	4.1		ns
C inputs via H2 through H	T_{HH2CK}	3.8		ns
C inputs via DIN	T_{DICK}	2.4		ns
C inputs via EC	T_{ECCK}	3.0		ns
C inputs via S/R, going low (inactive)	T_{RCK}	4.0		ns
C_{IN} input via F/G	T_{CCK}	2.1		ns
C_{IN} input via F/G and H	T_{CHCK}	3.5		ns

IOBs can be determined but the interconnects between them cannot (Figure 4.13).

4.5.8 Xilinx XC9500 CPLD

In addition to its families of FPGAs, Xilinx offers a family of CPLD devices fabricated in a CMOS flash process. The architecture of these devices evolved from PAL device architecture. They combine and interconnect a number of PAL-like blocks on a single chip. Logic within the blocks is implemented as wide AND-gates, ORed together, driving either a flip-flop or an output directly. They provide wide-input gating and system-clock rates of up to 150MHz, making them useful for implementing state machines and complex synchronous counters.

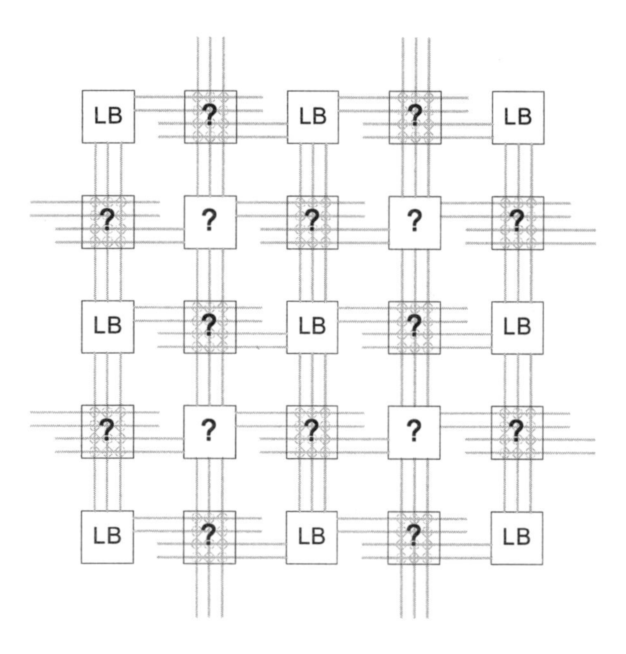

Fig. 4.13 The Routing Delays in Xilinx XC3000/4000 Devices Are Not Deterministic Prior to Placement and Routing

4.5.9 Xilinx XC9500 CPLD Architecture

The architectural features of the XC9500 family are shown in some detail in Figure 4.14. A device consists of a number of function blocks (FBs) and I/O blocks (IOBs) fully interconnected by a switch matrix. Connections through the switch matrix have a uniform, predictable delay. The number of function blocks varies from two for the XC9536 to 16 for the XC95288. A function block contains 18 macrocells, each capable of implementing a combinatorial or registered function.

Each function block contains logic to create combinatorial functions using sum-of-products implementations. A function block receives 36 inputs from the switch matrix and sends the true and complement version of each into a programmable AND-array to form 90 product terms (five per macrocell). The product term allocator allocates any number of these product terms to each macrocell by ORing the product terms of one macrocell with those of other macrocells. The concept of product-term sharing is shown in Figure 4.15. The additional time required for a signal to propagate through such logic depends on the number of allocators used to form the

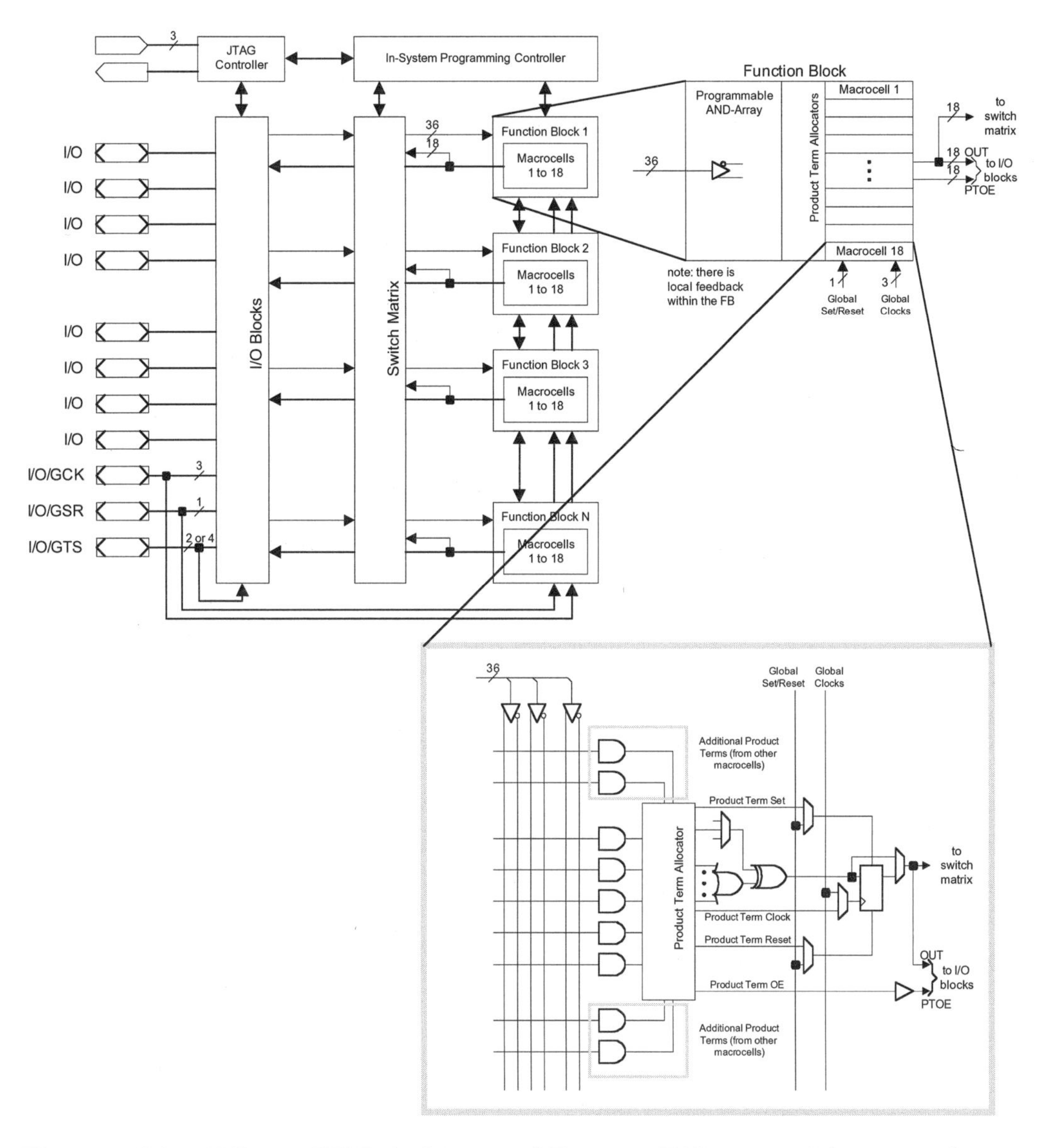

Fig. 4.14 Xilinx XC9500 CPLD Architectural Features [Xilinx, 1998] Courtesy of Xilinx, Inc. Copyright © Xilinx, Inc. 1999. All rights reserved.

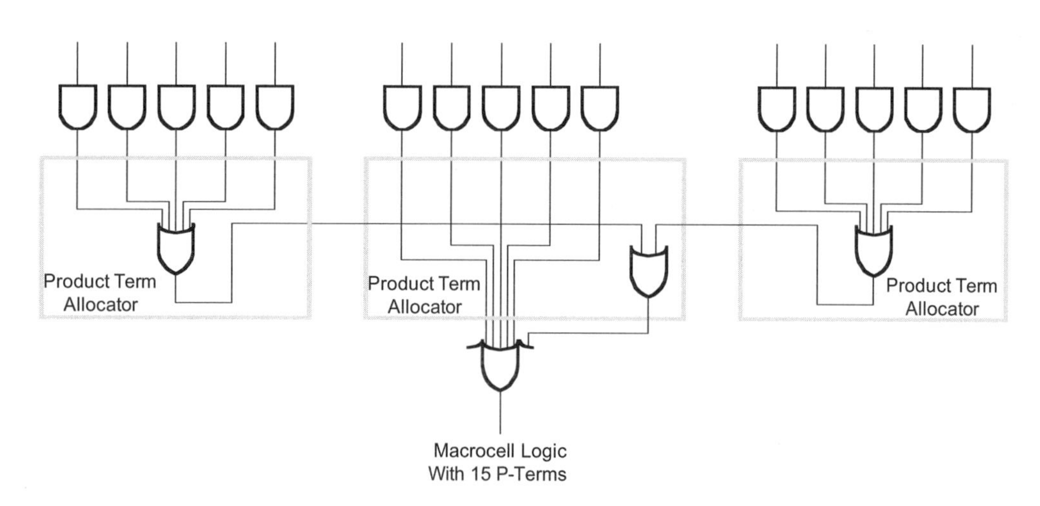

Fig. 4.15 Product Term Allocation with 15 Product Terms [Xilinx, 1998]

product term. Xilinx defines the term span (S) to be one less than the maximum number of allocators in the product term. The extra delay incurred equals S times the incremental product term allocator delay (t_{PTA}). The case illustrated in Figure 4.15 for 15 product terms results in a delay penalty of (2-1)*t_{PTA}.

One other feature of the architecture that affects timing is output-buffer slew-rate control. The edge rate of each output can be individually slowed down to reduce noise.

4.5.10 Xilinx XC9500 CPLD Timing Model

The structure of the XC9500 device allows the timing performance of a design to be gauged prior to completing all of the steps of the vendor's development process. To help the user predict timing, Xilinx provides the set of timing models shown in Figure 4.16. The expected performance of a device can be determined by using the timing parameters as defined by these timing models in conjunction with their actual values presented in the individual device data sheets (see Example 4.3). A partial listing of the timing parameters for the Xilinx XC95144 is shown in Table 4.10.

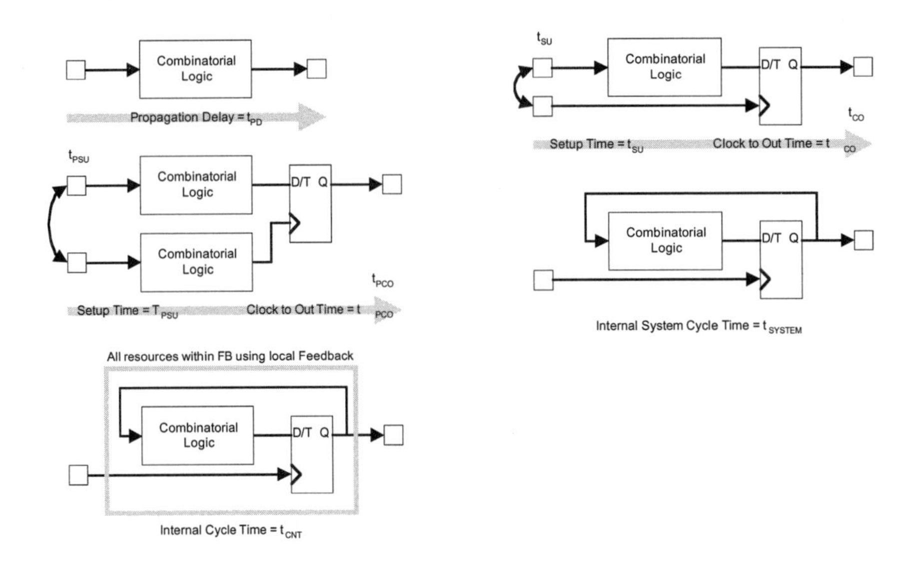

(a) Basic Timing Models

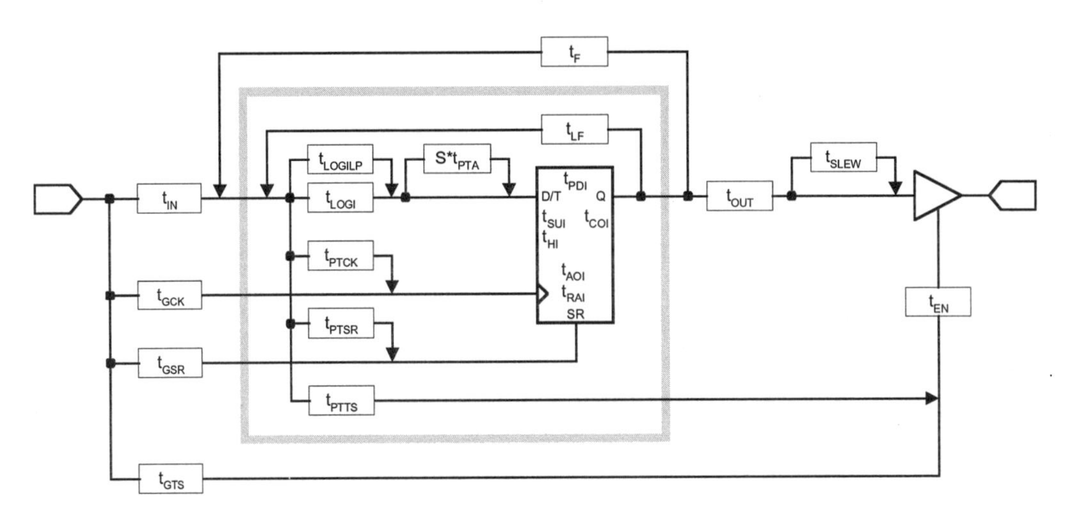

(b) Detailed Timing Model

Fig. 4.16 Xilinx XC9500 CPLD Timing Models [Xilinx, 1998] Courtesy of Xilinx, Inc.

Table 4.10 Xilinx XC95144 Timing Parameters (partial listing) [Xilinx, 1998]

		XC95144-7		
Symbol	**Parameter**	**Min**	**Max**	**Units**
AC Characteristics				
t_{PD}	I/O to output valid		7.5	ns
t_{SU}	I/O setup time before GCK	5.5		ns
t_H	I/O hold time after GCK	0.0		ns
t_{CO}	GCK to output valid		5.5	ns
.	.	.	.	.
.	.	.	.	.
.	.	.	.	.
Internal Timing Parameters				
T_{IN}	Input buffer delay		2.5	ns
t_{GCK}	GCK buffer delay		2.5	ns
t_{GSR}	GSR buffer delay		4.5	ns
t_{GTS}	GTS buffer delay		7.0	ns
.	.	.	.	.
.	.	.	.	.
.	.	.	.	.
t_{PTA}	Incremental product term-allocator delay		1.0	ns
t_{SLEW}	Slew-rate limited delay		4.0	ns

Timing parameters defined in the AC characteristics part of the table correspond to the basic timing model and are external or pin-to-pin timing parameters. Timing parameters defined in the internal timing parameters part of the table correspond to the detailed timing model. Only the columns for speed grade 7 are shown. This family contains the additional speed grades of 5, 6, 10, 15, and 20. Not all parts come in all speed grades as shown in Table 4.11. The Xilinx speed grade is tied to the t_{PD} timing parameter. The lower the speed grade the faster the device. The larger devices in the family do not come in the faster speed grades, but the timing parameters

Table 4.11 Xilinx XC9500 Family Performance Summary [Xilinx, 1998] Courtesy of Xilinx, Inc.

Device	Number of Macrocells	Speed Grades	Internal Operating Frequency (MHz)
XC9536	36	5, 6, 7, 10, 15	100
XC9572	72	7, 10, 15	83
XC95108	108	7, 10, 15, 20	83
XC95144	144	7, 10, 15	83
XC95216	216	10, 15, 20	67
XC95288	288	15, 20	56

are constant across all members of the family for a particular speed grade.

There are further complications to consider when using the Xilinx XC9500 timing models. The timing models, as shown, are valid for macrocell functions that use the direct product term only, with standard power setting and standard slew-rate setting. Several timing-parameter values must be modified when these conditions do not hold. Timing modifications are presented in the internal timing parameter tables as time adders. If they apply, they are added to the nominal value of the parameter.

Example 4.3

Consider how t_{PD} changes when three additional product term allocators are used to create a function (giving a span of two) and the output is slew limited. The values of the parameters are (for a XC95144-7 device):

$t_{PD} = 7.5ns$
$t_{PTA} = 1.0ns$
$t_{SLEW} = 4.0ns$

and,

$$t_{PD}(modified) = t_{PD} + S^*t_{PTA} + t_{SLEW}$$

Example 4.3 (Continued)

where S is equal to the span which is equal to (# of product term allocators) – 1
therefore,

$$t_{PD}(\text{modified}) = 7.5\text{ns} + 2*1.0\text{ns} + 4.0\text{ns} = 13.5\text{ns}.$$

4.6 HDL SYNTHESIS

Producing an efficient implementation of a design, when speed or area is critical, is most easily done by an experienced designer using schematic capture. As designs have become larger and larger this path may not be an option for a design and HDL synthesis must be used. Other reasons to use HDL synthesis for programmable logic design are:

- Once the source HDL code has been developed and tested, synthesis can target any vendor's device for which there is a synthesis library.
- Changes are easier to make, especially significant changes, in a design that has been captured in HDL code rather than in a schematic.
- The designer must use a company-approved design flow that requires HDL synthesis.

The principles of using HDL constructs and synthesis to generate gate-level logic structures, as applied to ASICs, have been covered in a previous chapter. These principles also apply to programmable logic design.

When using HDL synthesis a significant part of the design effort revolves around writing code that will produce a gate-level implementation of the design that satisfies the required timing and area constraints. In this regard, knowledge of the target device architecture is helpful, as is a thorough knowledge of how each HDL source code construct translates to a gate-level implementation. Information is also available from the programmable-logic-device vendors, and from the synthesis tool vendors. However, probably the most useful information is obtained through the use of small test cases, and trial synthesis runs on all or part of a design. Two recommendations are worth noting at this point:

- ☞ Use one-hot encoding for state machines
- ☞ Take advantage of register primitive gated clock logic

Synopsys (FPGA Compiler and FPGA Express), Exemplar, and Synplicity, among others, supply synthesis tools targeted at the field of programmable logic design. These tools contain optimization algorithms to allow the proper utilization of the architectural features specific to each supported programmable logic family. For example, the design flow for the synthesis of Altera FLEX family members using an Exemplar synthesis tool called Leonardo is shown in Figure 4.17. The Altera FLEX specific optimization algorithms are, as described by Exemplar [Exemplar, 1998]:

- ☞ **Fanin Limited Optimization Algorithm:** This algorithm utilizes the limited number of inputs of the logic element.
- ☞ **Look-Up-Table Mapping Algorithm:** This algorithm generates optimal mapping of combinatorial logic into look-up-tables (LUTs) and cascade gates.
- ☞ **Data-Path Synthesis:** This algorithm infers and implements data-path elements, such as counters, using Modgen to take advantage of the carry and cascade chains which are features of the FLEX architecture.
- ☞ **Clock-Enable Detection:** This algorithm automatically detects register clock-enable logic
- ☞ **Constraint-Driven Timing Optimization:** This algorithm understands the FLEX architecture and speeds up critical paths in the design.

The Exemplar Leonardo Modgen feature invokes a library of predefined gate-level, technology-specific implementations of common data-path elements such as counters and adders. When using Modgen, the designer can specify the degree to which speed (at the expense of area) is important via a synthesis variable. Synopsys provides a similar feature called XBLOX.

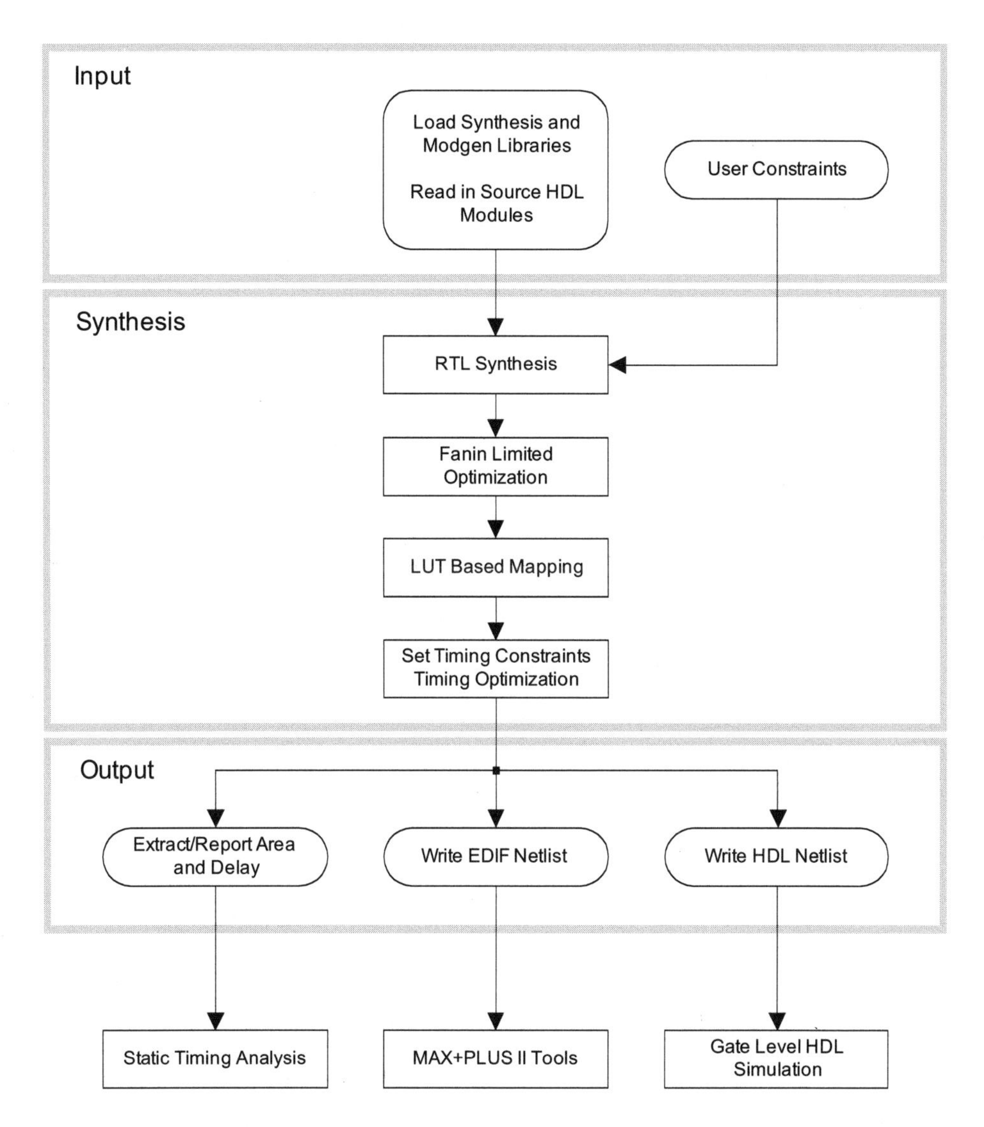

Fig. 4.17 Synthesis Design Flow (Exemplar Leonardo to Altera MAX+PLUS II) [Exemplar, 1998] Courtesy of Exemplar logic.

4.7 SOFTWARE DEVELOPMENT SYSTEMS

The vendor-specific software, or development system, used to create the physical realization of a programmable logic device consists of the following elements:

- Libraries for schematic capture, HDL simulators, and synthesis tools
- Compiler to map the design to vendor-specific cells
- Place-and-route software
- Post/place-and-route delay extraction and SDF file creation
- Static-timing analysis tools
- Programming file creation software

After a design is captured and verified at the functional level a netlist is generated. The netlist, which oftentimes is in EDIF format, is input to the vendor-specific tools. In most cases the design must meet certain timing-performance goals and the place-and-route process must be guided by placing constraints on timing. The constraints are conveyed to the place-and-route software in a variety of ways. They may be entered as properties on a schematic, be provided in a separate constraint file, or be entered interactively using the development system's graphical user interface (GUI).

Until placement and routing is completed, timing has only been roughly estimated because, in most cases, actual device timing is very much dependent on routing delays. Following place and route, timing analysis is performed. The development system reports how well it met the timing constraints. Full functionality static-timing-analysis tools are usually part of the development system. They generate a detailed path-by-path analysis of delays. Most development systems can also extract postplace-and-route delay information and create an SDF file. The SDF file can be used to back annotate delays into gate-level HDL files, allowing resimulation with actual delays. The device's dynamic timing can be verified by simulating the back-annotated design; however, only those paths exercised by the functional simulation vectors are verified.

4.7.1 Timing Constraints

Properly defining and applying timing constraints are key elements in the design process. Realistic constraints lead to successfully achieving timing-performance goals. Constraints need to be set intelligently because the act of setting a constraint does not guarantee the device will be able to meet the constraint. The place-and-route software does its best with what it has. It is possible to negatively impact results with improper constraint definition because of limited routing resources. For example, the following are points to consider, as recommended by Actel in its on-line help:

☞ **Set Sufficient Constraints:** All necessary constraints for a design need to be defined to ensure the desired result. Undefined paths have a low priority for resource allocation.

☞ **Avoid Unnecessary Constraints:** Paths that are actually don't-care paths should not be constrained. If they are, they will use up resources that high-performance paths need and which may be in short supply.

☞ **Avoid Overconstraining:** Constraints should not be defined too short. Defining a constraint to be shorter than what is actually required for margin can negatively impact performance because of competition for device resources.

Example 4.4

Actel provides the following as an example of overconstraining a 40MHz clock (25ns period) design:

Defined Constraint	*Postplace-and-Route Result*
25.0ns	24.8ns
24.0ns	23.9ns
23.0ns	24.5ns
22.0ns	25.2ns

The device runs out of overall resources at 23.0ns, making the performance goal impossible to reach. Further constraining the device only causes its performance to degrade further.

Timing constraints can be grouped into the following two categories:

☞ **Clock Constraints (register to register):** Clock constraints define the timing of the clock networks. The period and duty cycle are defined for each clock network or domain. Timing for the register-to-register paths associated with each clock domain is constrained accordingly. Register-to-register paths that are not to hold to this timing must be explicitly defined.

☞ **Path Constraints:** Path constraints define the delays from input pad to output pad, input pad to register, and register to output pad. These constraints can usually be set on a class basis so that each path does not have to be explicitly defined. For example, all register-to-output pad paths can be constrained to have a particular delay value with one command.

4.7.2 Operating Conditions

For timing-analysis results to have any meaning, realistic operating conditions must be determined and fed into the place-and-route software. Setting the operating point is usually done interactively as part of the development system's graphical user interface. The device's speed grade must also be selected at the same time. The operating conditions consist of process, temperature, and voltage. Process variation and speed grade are vendor specific. The decision of which speed grade to select is driven by performance goals and affects the ultimate cost of the product of which the device is a part. Temperature is a function of the device's power consumption and the environment in which it is operating.

4.7.3 Static Timing Analysis

For the most part, programmable logic vendors provide full-function, static-timing-analysis tools similar to those used for ASIC development. STA tools use the extracted postroute-delay information along with the constraint information to provide a path-by-path analysis of timing delays. The analyzer's operation can usually be directed from a graphical user interface or from the command line, and can present its results in a number of different ways. The types of information it provides are:

- ☞ Register-to-register delays
- ☞ Clock-to-output pad delays
- ☞ Input pad-to-output pad delays
- ☞ Input pad-to-register delays

A static timing analyzer can provide delay data for all paths or it can be directed to present data for only those paths that have violated their timing constraints. It can be further directed to report only a selected number of the worst violators.

4.7.4 Vendor-Specific Timing-Verification Tools

The following sections briefly detail the timing verification facility specifics of software development systems from Actel, Altera, and Xilinx. A summary of the timing-verification facilities contained in each of these development systems is presented in Table 4.12.

Table 4.12 Summary of Vendor Timing-Verification Facilities

Timing Verification Function	**Vendor Development System**		
	Actel Designer	**Altera MAX+PLUS II**	**Xilinx XACT/M1**
Define Timing Constraints	DirectTime Edit (GUI) Constraints File	Global Project Timing Requirements	Enter attributes on schematic Constraints File
Static Timing Analysis	DirectTime Analyzer Timing Report Generator	Timing Analyzer	Timing Analyzer
Dynamic Timing Analysis	SDF Back-Annotation	SDF Back-Annotation MAX+PLUS II Simulator	SDF Back-Annotation
Comments	Two types of layout supported: Standard and Timing Driven (DirectTime Layout)		

4.7.5 Actel Designer

Actel provides two design paths for their FPGA devices, standard and timing driven. They call their timing-driven path DirectTime layout (DTL). A timing-driven or DirectTime flow diagram is shown in Figure 4.18. The Actel timing-verification tools consist of DirectTime edit, DirectTime layout, and DirectTime analyzer. Actel suggests that a quick, prelayout timing estimate of the design be done using either the DirectTime analyzer or the timing report generator

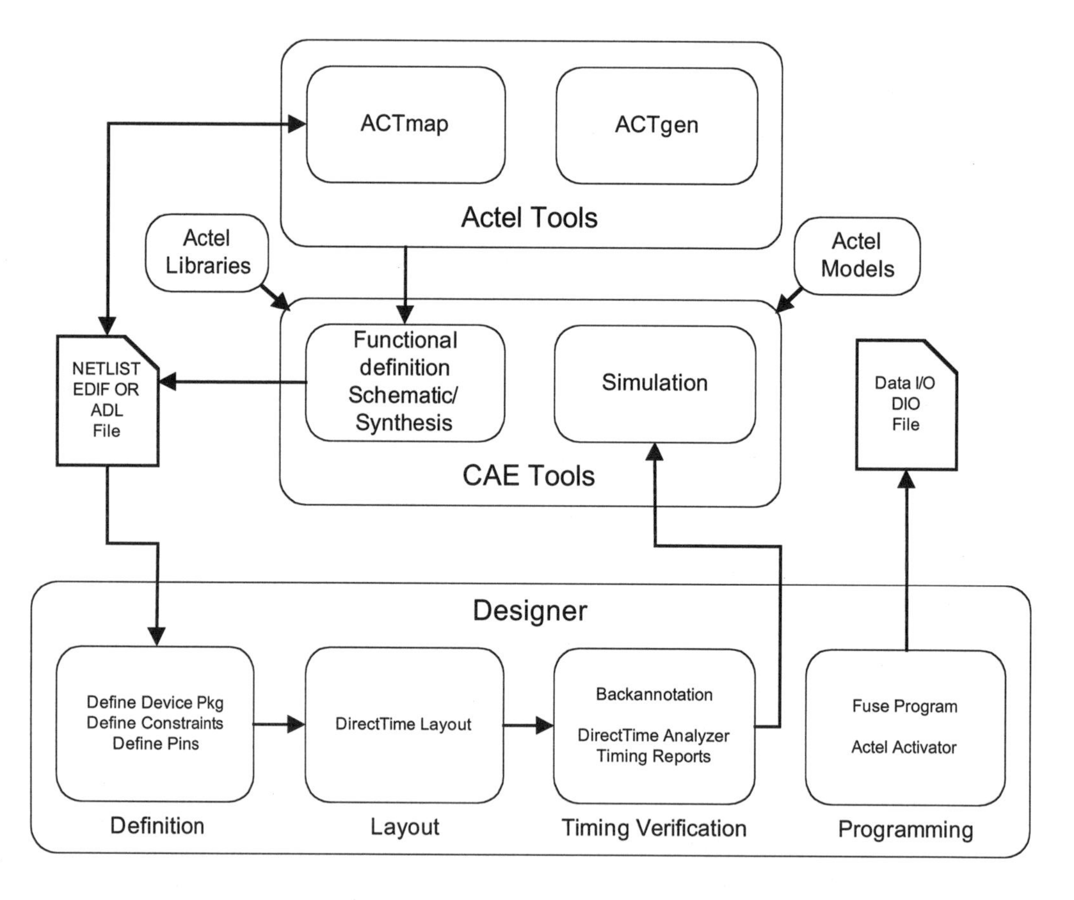

Fig. 4.18 DirectTime Flow [Actel98] Courtesy of Actel Corporation

(see below). If timing is not within ±15% of the design timing requirements then the DirectTime layout path should be used.

DirectTime (DT) edit is a graphical interface for defining delay constraints to control the DirectTime layout engine. Constraints may also be defined in an external file. The DTL engine considers the defined delays when allocating silicon resources with the goal of meeting or beating all constraints if possible. The DTL engine tries to meet the constraints by considering the performance criticality of one function versus another when allocating device resources.

The DirectTime (DT) analyzer analyzes static timing on a DirectTime layout terminal basis. (Layout terminal refers to the starting and ending points for a signal path.) It is an interactive tool whose results depend on the previously defined operating conditions. The DT analyzer can use either prelayout estimated delays or postlayout calculated delays, based on the selected device and defined operating conditions. Used in conjunction with the DirectTime editor and DirectTime layout, it displays delay-constraint goals versus actual results for any specific path or clock constraint.

DT's analyzer presents a summary of its results in a tabular format. Any path in the results list can be expanded to show the individual logic-component delays that contribute to the total path delay (called an expanded list). One additional feature, the expanded graphical chart, is useful in understanding the complexity of a given timing problem.

The DT analyzer is intended to be an interactive diagnostic tool to help debug specific timing problems. The timing report generator should first be used to determine if any such problems exist in a design. The default timing report will automatically contain the following information:

☞ Maximum delay from input I/O to output I/O
☞ Maximum delay from input I/O to internal registers
☞ Maximum delay from internal registers to output I/O
☞ Maximum delays for each clock network
☞ Maximum delays for interactions between clock networks

4.7.6 Altera MAX+PLUS II

The Altera MAX+PLUS II design environment, shown in Figure 4.19, includes the MAX+PLUS II timing analyzer.

Timing requirements for a project can be entered at the global level to specify the overall requirements for:

- Input-to-nonregistered output delays (t_{PD})
- Clock-to-output delays (t_{CO})

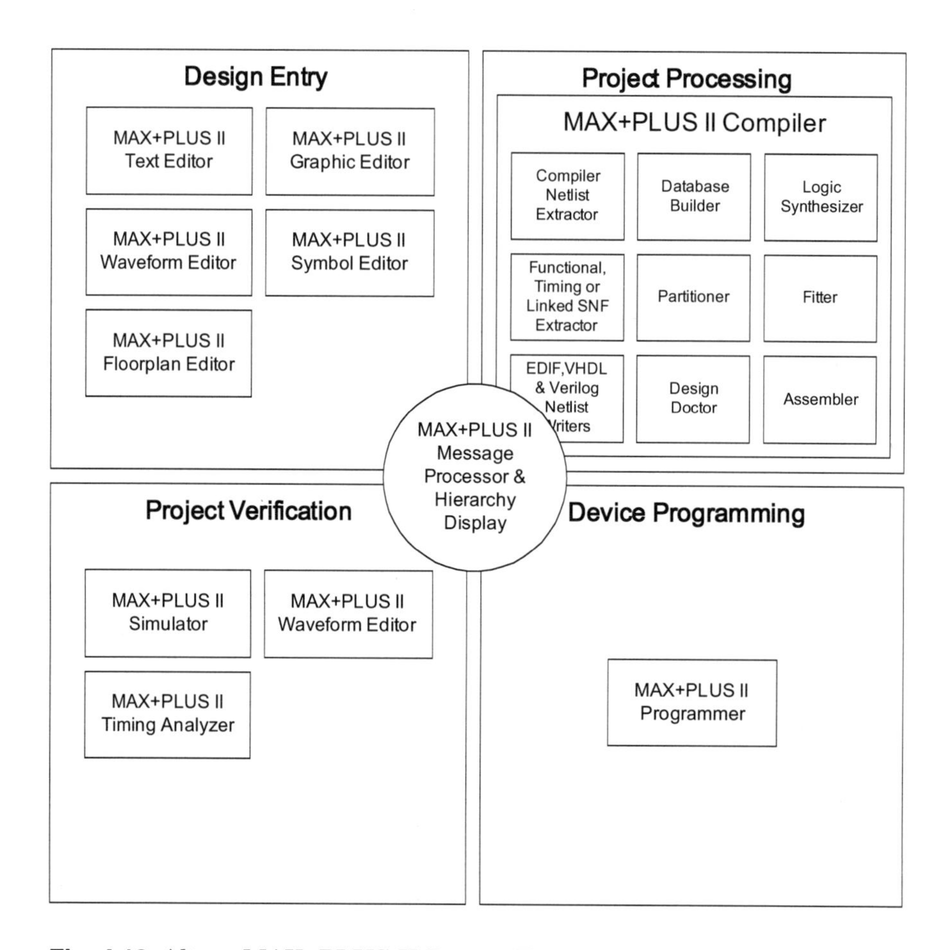

Fig. 4.19 Altera MAX+PLUS II Design Environment [Altera Corporation, Sept. 1997] Courtesy of Altera Corporation.

- Clock setup time (t_{SU})
- Clock frequency (f_{MAX})

Also, connections between all bidirectional-feedback, preset-signal, and clear-signal timing paths can be cut.

The MAX+PLUS II timing analyzer is used to analyze the timing performance of a device after the compiler has optimized it. The compiler creates a timing simulator netlist file (.snf), after the device is fully synthesized and optimized. All signal paths can be traced, and failing critical speed paths determined.

The timing analyzer generates three types of analyses:

- The delay matrix shows the shortest and longest propagation delay paths between multiple source and destination nodes in a device.
- The setup/hold matrix shows the minimum required setup and hold times from input pins to the data, clock, clock enable, latch enable, and address, and from write enable inputs to flip-flops, latches, and asynchronous RAM.
- The registered performance display shows the results of a registered performance analysis. It analyzes registered logic and determines minimum clock period and maximum circuit frequency.

Nodes can be tagged as sources and destinations for timing analysis to provide point-to-point propagation delays, and setup- and hold-time requirements. The maximum clock frequency for each clock signal in a design can also be calculated.

4.7.7 Xilinx XACT/M1

The Xilinx timing analyzer performs static timing analysis of an FPGA or CPLD design. Before timing analysis can be done an FPGA design must be mapped and can be partially or completely placed and routed. A CPLD design must be completely placed and routed. The Xilinx timing analyzer does not perform setup and hold checks. Dynamic timing verification using functional simulation must be used to verify setup and hold times. Figure 4.20 shows how timing verification fits into the Xilinx design flow.

Timing constraints can be applied to a net or path that is identified by specifying its start and end points. Groups can be defined whose start and end points are flip-flops, I/O pads, latches, or RAMs. A single delay requirement for all paths in the group can be

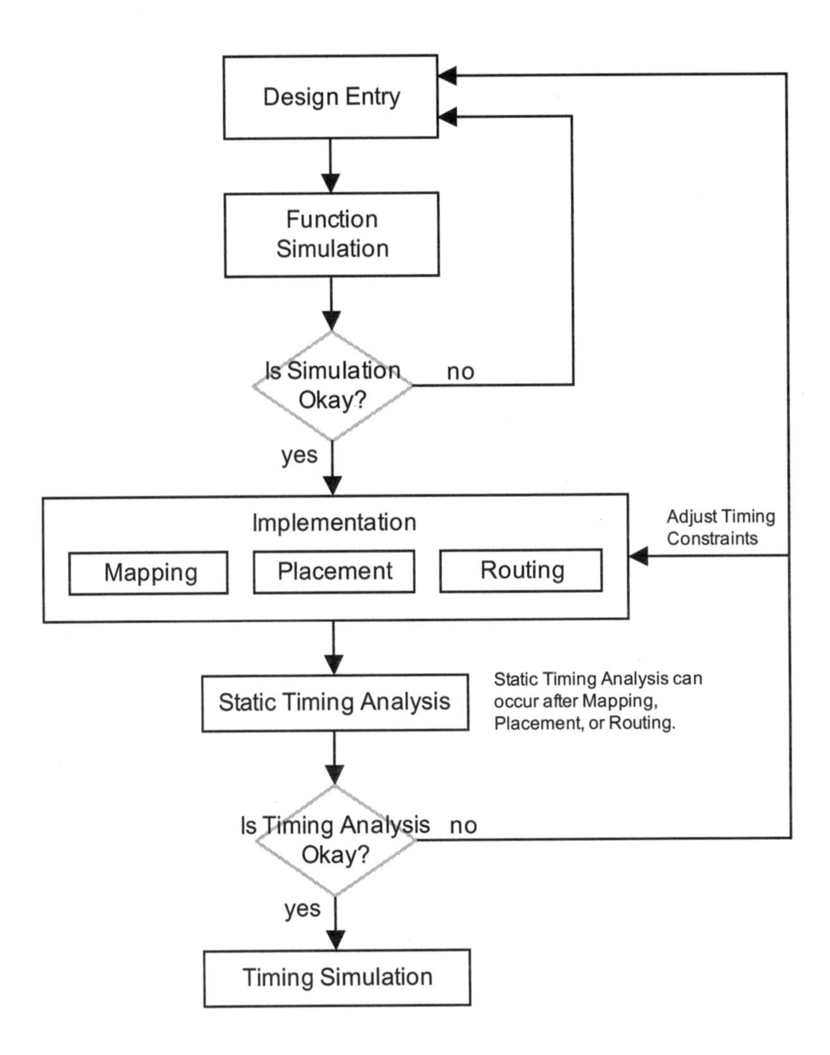

Fig. 4.20 Timing Analyzer in the Xilinx Design Flow [Xilinx, 1999] Courtesy of Xilinx, Inc.

specified. Constraints can be entered as attributes/properties on a schematic or as attributes in HDL source-code files. They can also be specified in a constraint file (i.e., a file with a .pcf extension).

Xilinx provides a graphical user interface to the timing analyzer. Commands can also be activated from the command line or by running macros that contain command sequences. The analyzer outputs results in the form of reports.

- ☞ The timing constraints analysis report compares design performance to the timing constraints.
- ☞ The content of the advanced design analysis report varies, depending on whether the design is for an FPGA or CPLD.

When the design is for an FPGA, the report displays the results of analyzing the constraints specified in the constraints file. If no constraints are specified, the report displays the maximum clock frequencies for all clocks in the design and the worst-case timing for all clock paths.

When the design is for a CPLD, the report displays all external synchronous path delays which include: pad-to-pad (t_{PD}), clock-pad-to-output-pad (t_{CO}), setup-to-clock-at-the-pad (t_{SU}), and internal clock-to-setup (t_{CYC}) paths.

- ☞ The custom analysis report contains a detailed analysis of all specified paths and includes the worst-case path delays for all paths in the design.
- ☞ The query nets report displays net delay information for FPGAs.

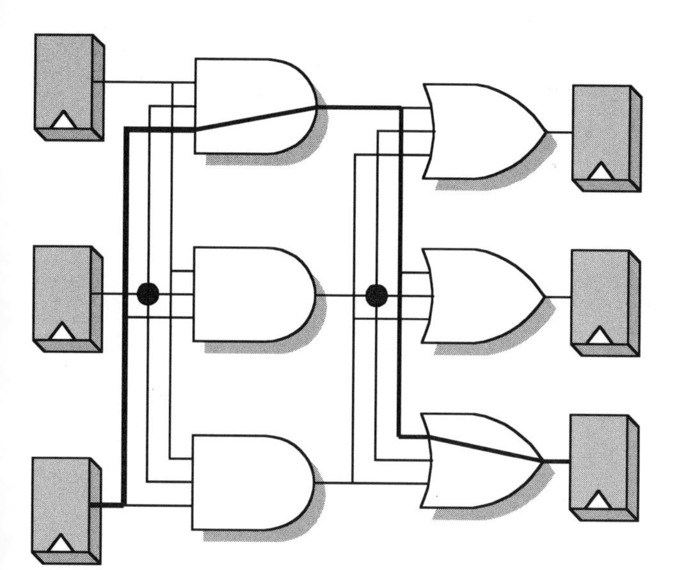

A

PrimeTime

PrimeTime is a chip-level, standalone static timing analyzer from Synopsys. It provides comprehensive reports on timing violations such as:

- Setup and hold requirements of sequential devices and gated clocks
- Minimum period and pulse width for clock signals
- Recovery and removal checks
- User-defined minimum and maximum delay constraints
- Maximum skew

The mentioned timing constraints are checked based on maximum and minimum case industrial-grade device conditions. PrimeTime provides accurate timing analysis through support of designs with gated clocks, multiple clocks and phases, multiple functional modes, multicycle paths, and false-path detection.

PrimeTime's modeling capabilities support the analysis for both top-down and bottom-up design approaches for the entire chip, not just the synthesized portions. PrimeTime's modeling techniques are:

- Extracted model
- STAMP model
- Quick timing model (QTM)

Extracted models are the most accurate models that can be used in PrimeTime. They are context-independent timing models from gate-level netlists and can be extracted under multiple operating conditions. The timing model extraction feature in PrimeTime increases designer productivity by reducing analysis runtime and memory usage in a bottom-up design flow. The extraction capability can also be used by core providers to supply their customers with accurate timing models. This approach protects their intellectual property because the extracted timing models do not depend on the design environment.

STAMP models are static timing models for complex blocks such as DSPs and RAMs. STAMP models are created by core or technology vendors who provide database (db) files for their customers as timing models. These models are usually generated for transistor-level designs where there is no gate-level netlist. STAMP models contain pin-to-pin timing arcs, setup, hold, mode, pin capacitance, and derive information.

Quick timing models are used to quickly create estimated timing models for unfinished blocks. This allows the designer to perform full-chip timing analysis by removing the warnings caused by existing black boxes in a design. These models contain information such as: clock definitions, delays, input/output pin list, load, and capacitance. Quick timing models should eventually be replaced by gate-level netlists as they become available.

PrimeTime is synthesis compatible and can fit into an existing synthesis-based design flow using standard Synopsys technology files for cell data, so no new libraries have to be created. PrimeTime reads Verilog, VHDL, EDIF netlists, Synopsys-format design databases, standard delay format (SDF), standard parasitic format (SPEF), Cadence Design Systems reduced standard parasitic format (RSPF), and detailed standard parasitic format (DSPF) files. Figure A.1 shows the PrimeTime input and output files.

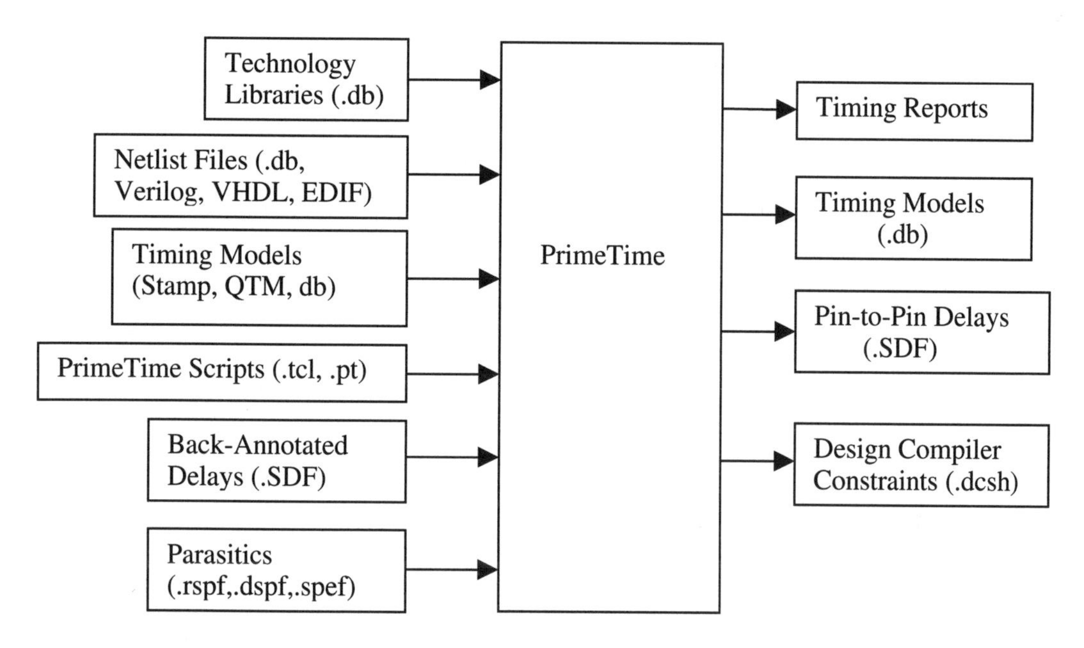

Fig. A.1 PrimeTime's Input/Output Files

PrimeTime can read the input files shown in Fig. A.1 by using the GUI command Open File or by the command line read_[format] [filename].

A summary of the basic steps to run PrimeTime is represented in Figure A.2.

The steps shown in Fig. A.2 can be implemented in PrimeTime using the graphical user interface or can be written as a PrimeTime script (.pt) for automating the process. The commands entered from the command line or GUI are saved in a log file and can be used as PrimeTime scripts with minor modifications for later use.

PrimeTime is compatible with and consistent to Design Compiler (i.e., Synopsys's synthesis tool). They share the same delay calculation subsystems and produce identical delay values. PrimeTime and Design Compiler use the same libraries and support the same .db format files.

PrimeTime accepts outputs from other synthesis tools such as Exemplar's Leonardo-Spectrum and AMBIT's BUILDGATES. Fig-

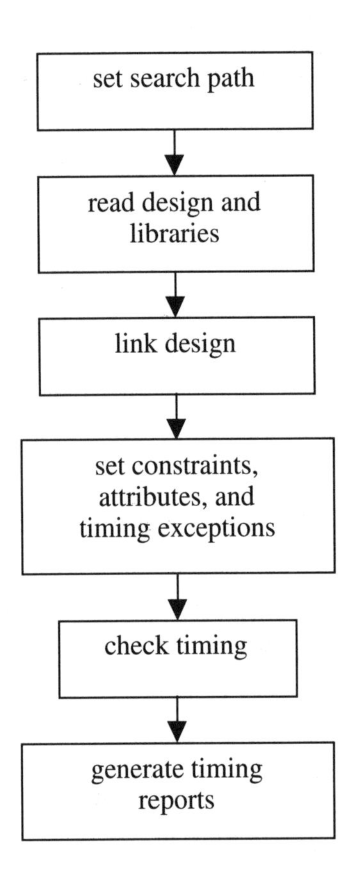

Fig. A.2 PrimeTime User Flow

ures A.3 and A.4 represent the input and output formats from these two synthesis tools.

There are two versions of Leonardo-Spectrum, called Level 2 and Level 3. The Level-2 version is used only for FPGAs; the Level-3 version is used for ASICs.

Both of the mentioned synthesis tools generate formats that can be used in PrimeTime. The only format they are not able to generate is db format which can be generated from the Design Compiler. It should be mentioned that BUILDGATES and Leonardo-Spectrum can perform the STA on a design, eliminating the use of PrimeTime for their outputs.

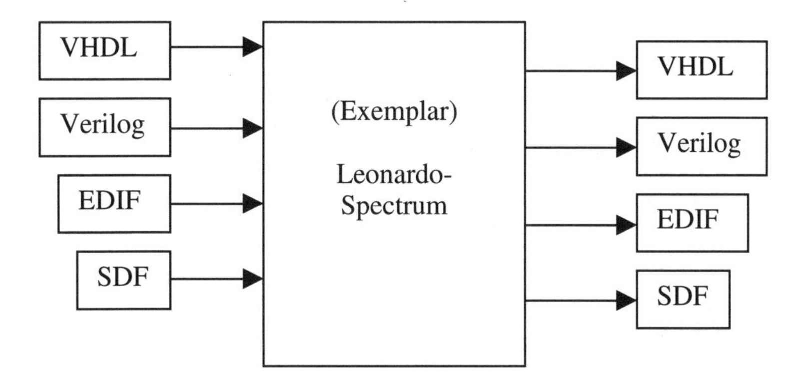

Fig. A.3 Exemplar Synthesis Tool Input/Outputs

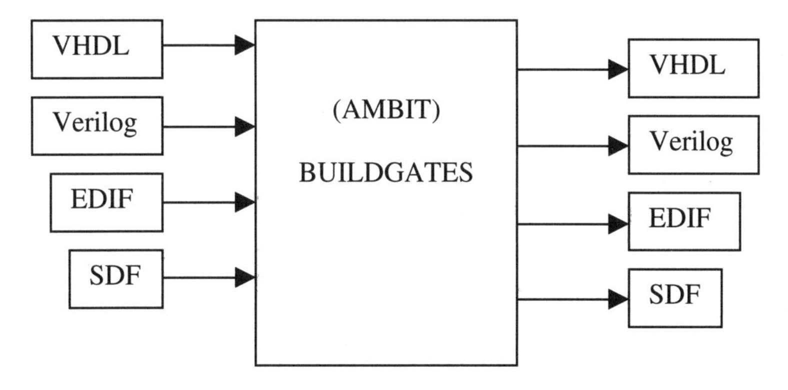

Fig. A.4 AMBIT Synthesis Tool Input/Outputs

Although PrimeTime is a powerful tool for static timing analysis (STA) of ASICs, it is not targeted for board-level or system-level static timing analysis. The reason is that some additional types of analysis which are only applicable at board-level, such as crosstalk analysis and signal integrity, do not exist in PrimeTime. However, the STA for board-level and multiboard-level designs can be done by BLAST (board-level application for static timing) from Viewlogic System Inc. BLAST is based on MOTIVE for board-level analysis with all the appropriate interfaces. It can read the STAMP models from PrimeTime. Therefore, PrimeTime can perform STA for ASICs and BLAST can perform the analysis for multiple ASICs (modeled in STAMP) in addition to other discrete components.

For more information on PrimeTime, contact:

Synopsys Corporate Headquarters
700 East Middlefield Rd.
Mountain View, CA 94043
(650) 962-5000
www.synopsys.com

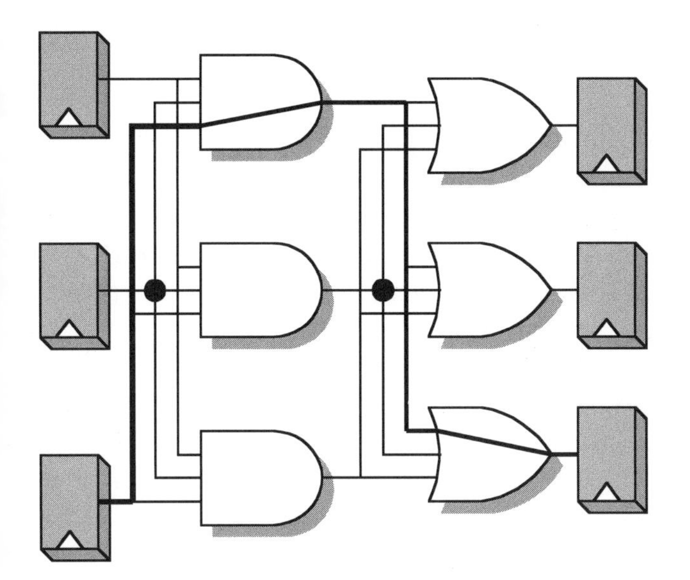

B

Pearl

Pearl is the static timing analyzer from Cadence Design Systems, Inc. Pearl identifies critical paths in combinational or clocked circuits, for individual blocks or the entire chip.

As a static timing analyzer, Pearl does not require simulation vectors, which allows timing analysis to start as soon as the netlist becomes available. Moreover, unlike a simulator that only reacts to input vector stimuli, Pearl traces and analyzes all paths in the circuit and reports longest paths and timing violations. Since vector-based timing simulation is slow and only as good as the vectors supplied, Pearl's faster run times and thorough exercise of paths significantly improve the reliability of timing verification.

Pearl shortens the timing-verification process by employing unprecedented core technology, dramatically accelerating what typically has been a critical but painstaking and time-consuming task. With previous-generation timing analyzers, timing verification

---

This appendix is courtesy of Cadence Design Systems, Inc. Portions reprinted with permission.

barely fit in the design schedule, despite the potentially devastating consequences of letting a few critical paths go unchecked.

Pearl readily accepts industry-standard netlist formats, including Verilog, Hspice, and Lsim, eliminating preprocessing so that analysis can begin as soon as the netlist is ready. In addition, Pearl uses smart-search algorithms to reduce path searching time.

Pearl's accuracy is within 15% of SPICE, accepting both back-annotated resistance and capacitance for accurate analysis. Pearl translates the circuit network into stages consisting of resistor and capacitor trees. This formulation avoids classic problems in analyzing pass transistors and latches. To ensure accuracy, the delay model depends on gate-input slew rate, as well as interconnect resistance and capacitance.

Pearl provides both transistor- and gate-level modeling and produces precise results with both. The transistor model is dependent on the input slew rate to ensure accurate delay calculation. Using circuit patterns supplied by the user to identify transistor-level latch and register structures, Pearl automatically determines setup- and hold-time constraints for these patterns. Pattern matching allows Pearl to work on a large variety of circuit structures rather than a few built-in structures.

To accelerate the analysis and support of today's structured custom design methodologies, Pearl also supports gate-level models ranging from simple inverters to complex RAMs, PLAs, and megacells. Timing models consist of pin-to-pin delays and setup- and hold-time requirements.

With an extensive selection of analysis commands to effectively identify and analyze chip timing, Pearl finds the critical paths in combinational logic, locates setup- and hold-time violations in clocked circuits, and is capable of handling circuits with multiple overlapping or nonoverlapping clock phases. Pearl even finds the minimum cycle time to monitor progress while tuning the design. Furthermore, Pearl can identify nodes with slow rise or fall times, long paths from a node, or paths between two nodes that serve as pointers to hazardous paths.

Pearl's variety of reports help designers examine the circuit network and delay calculations. For example, detailed path-trace reports can be generated interactively for quick feedback; or, to run SPICE on a path, Pearl can generate a complete SPICE deck for those specific transistors on the critical path, which significantly reduces the time required for circuit simulation. Pearl also supports several features to suppress and filter false-path identification, helping to reduce the amount of information reported to the user, and ultimately making it easier to identify and correct the real timing problems.

With today's design complexities and competitive markets, project teams will often use a team design strategy to expedite design completion. Pearl's mixed-mode, hierarchical timing verification supports this divide-and-conquer approach so that groups progressing at different rates can verify full-chip performance without stretching the project deadline. Further, Pearl's support of mixed-level design lets designers model unfinished blocks with timing models that correspond to their target timing specification. The models can then be verified with others for full-chip analysis. Alternatively, timing models can be used in place of a cell's transistor netlist to speed up the analysis process. Pearl lets designers verify their design in the most appropriate fashion throughout the design cycle.

With the timing effects inherent in today's deep-submicron geometries, timing analysis must play a central role in a cohesive timing-driven design flow. Pearl's ability to look out through I/O ports to generate arrival times, input drive strengths, and output loads and constraints lets designers generate synthesis constraints for a top-down, timing-driven methodology. As a design progresses, Pearl reads delays in SDF format from a floorplanned, placed, or fully routed design to check that the design is converging on its performance goals. At the end of the design process, Pearl reads back-annotated resistance and capacitance values from DRACULA extraction as a final check to assure that the design meets timing requirements.

Silicon that is functionally correct is only half the story—it must also meet performance requirements to be viable in the marketplace. The cost of a design that runs slowly can be enormous due to the cost of spinning the silicon, but missing a market window can be even more costly. Investing in Pearl helps designers get their product out the door on time, as well as improve its performance.

Pearl's key features include:

- ☞ Static timing analyzer which reports critical paths and timing violations without the need for input vectors
- ☞ Mixed transistor- and gate-level analysis for accuracy and speed
- ☞ Full-chip analysis—eliminating the need to glue together partial results
- ☞ High-capacity Sun UltraSPARC (i.e., 5.2 million transistors, 167MHz)
- ☞ Speed—allowing iterations to improve chip performance (not just a final check at the end)
- ☞ Accuracy to within 15% of SPICE
- ☞ Extensive selection of analysis commands for work on a wide variety of circuits
- ☞ SPICE deck of critical path to reduce circuit simulation time
- ☞ Acceptance of back-annotated resistance and capacitance or delays in SDF format

For more information on Pearl, contact Cadence Design Systems, Inc., at www.cadence.com.

APPENDIX C

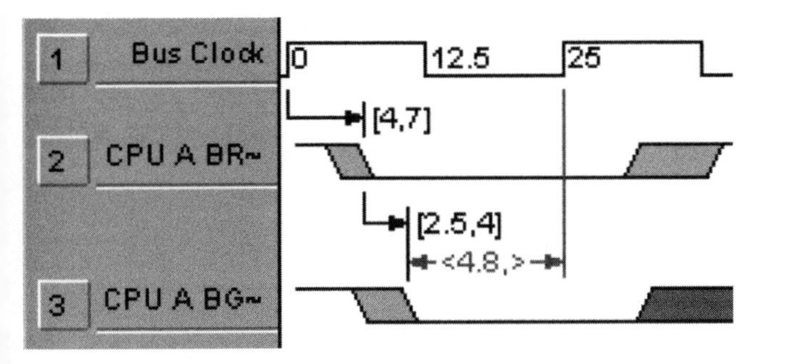

TimingDesigner

TimingDesigner Professional is an interface design tool which allows the designer to specify and analyze complex circuit interfaces up front in the design process. The resulting interface specification can then be used to accurately communicate design details to other people, teams, and design tools.

The key to interface design is to start with an intelligent timing diagram, which is an unambiguous electronic specification incorporating such constructs as signals, clocks, delays, constraints, samples, conditions, cycles, guarantees, grids, skews, variables, and text annotations.

Once there is an intelligent timing diagram, it can be used to answer critical design questions even before a netlist is produced, while there is still time to make meaningful changes.

For more information on TimingDesigner contact:

Chronology Corporation
Corporate Headquarters
14715 NE 95th St.
Redmond, WA 98052

(425) 869-4227 phone
(800) 800-6494 toll-free
(425) 869-4229 fax
www.chronology.com.

This appendix is courtesy of Chronology Corporation. Portions reprinted with permission.

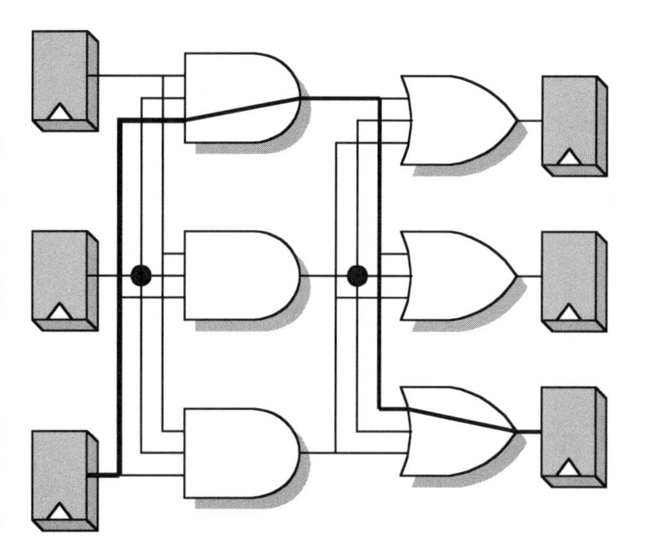

D Transistor-Level Timing Verification

The basic building block in silicon processing is the transistor. Digital simulation and synthesis tools are based on Boolean logic blocks, not transistors. Vendors build logic gates and storage elements out of the transistors and offer them in what is often called a digital cell library. Analog simulators are used to characterize these library cells and reduce accurate transistor-level models of the cell into a more manageable digital description of the cell's behavior. This simplification results in a model that can be simulated 250+ times faster.

Most modern digital systems follow synchronous logic design practices, with domino logic being one obvious exception. Synchronous logic design simplifies simulation, synthesis, and timing verification by restricting the timing design to a more easily solved static timing problem. Digital simulators and static timing analyzers (STA) utilize timing arcs to describe the delay from each input to each output. Timing arcs can also be used to describe constraints between input pins, such as setup and hold, or minimum pulse width. A library vendor will typically perform transistor-level simulations on each cell in the library and specify a timing triplet (minimum, typical, maximum) for each arc which covers a given range of temperature, process, and voltage. This was a reasonable simplification for 50MHz, 1 micron systems, but in today's 500MHz+, 0.15 micron designs some of the basic assumptions are no longer valid.

Table D.1 shows the transistor-level timing tools versus other forms of timing verification.

- ☞ **Library Format:** >1 micron transistors were modeled using foundry-specific algorithms. Most foundries have standardized on public domain BSIM3 for deep-submicron MOS transistors since it accurately models the subthreshold region. Timing library format (TLF) is a Cadence-specific library. A file with an .lib extension indicates a Synopsys-specific library.
- ☞ **Static/Dynamic:** Dynamic timing verification takes into account all effects including input load dependencies on output state. Static timing verification precalculates all delays. Newer STAs also precalculate signal slew rates for use in intrinsic delay calculations.
- ☞ **Logic Model:** Logic model is the basic model used by the timing analyzer.
- ☞ **Wire Model:** Distributed wire models allow for multinode spice netlists to be back-annotated in place of a single signal to

Table D.1 Table of EDA Timing Tool

Tool	Vendor	Library Format	Static/ Dynamic	Logic Model	Wire Model	Rel Speed	Rel Accuracy
Spice		Foundry Specific	Dynamic	Transistor	Distributed RLC Tree	1	98%
HSpice	Avant!	BSIM3V3	Dynamic	Transistor, some Gate	Distributed RLC Tree	4	100%
Pearl	Cadence	TLF	Static, with input slew	Gate, simplified Transistor	Distributed RC Tree	800	85%
PathMill	Synopsys	.lib	Static, with clock slew	Gate, simplified Transistor	Lumped	1000	80%–90%
STA			Static	Gate	Lumped	500–1500	50%–80%

approximate transmission line effects. Lumped models (like SDF) allow for a resistance and capacitance pair to be back-annotated onto the signals. Data to be back-annotated is extracted from the physical layout by layout tools.

D.1 HSPICE

Most silicon foundries provide BSIM3V3 (Hspice level 47) transistor models for their modern deep-submicron processes. BSIM3 is a public-domain set of equations, developed by the University of California at Berkeley, which accurately describes the behavior of deep-submicron MOS transistors. This replaces the many custom transistor models each vendor developed, each with its own set of inadequacies.

Accurate characterizations can be run against the transistor netlist of either logic gates or small subsystems (<4k gates). For the best accuracy the netlist should include parasitics extracted from the physical layout of the circuit. Using the analog simulator to characterize larger black boxes enables the paths inside the box to be very accurately modeled. During this simulation, data is collected for the interface pins of the black box. This data is then used to generate timing-arc triplets the same way as is done for the gate-level cells.

Steps to characterize a black box:

1. Specify which parameters (e.g., timing arcs, output drive, constraints) are needed by the target STA and the number of units for each. Specify if the parameters are a single value or a piecewise linear curve.
2. Specify the operating triplets (i.e., min, typ, max) to run for temperature, process, voltage.
3. Specify the input-drive and output-load conditions.
4. Specify a simulation stimulus (i.e., set of vectors) to exercise each arc to be characterized.

5. Run the vectors and collect the data.
6. Reduce the data and format into a library for the target STA.
7. Verify the STA library created.

Hspice can use one netlist to drive multiple simulations by specifying user-defined parameters which are altered between simulator runs. Typically, parameters would be defined for the temperature, supply voltage, process models, operating points, and output loads. The alter statement indicates the beginning of a new simulation set. Measure statements are used to extract the data from the analog waveforms; the data is then used to build the black box digital model.

Example D.1 (Spice Netlist for Characterization)

```
First run using slow models
* The first line is the title of the first simulation run
* Comments have * in first column
* A + in the first column is a continuation line
.global VDD
* some simulator directives to improve convergence and accuracy
.option post=1 NoMod Accurate=1 method=gear reltol=1e-5 kcltest=1
* set up signal logging for ports of black box
.option probe
.probe v(inclk) v(clk) v(data) v(ena) v(zen)
* do not include protected items in transcript
.protect
.LIB '/libraries/models.sp' "slow"
.inc '/libraries/gates.sp'
.unprotect
* define a macro for the pads & wire bonds
.macro pin
+ pi po lpin=10.2n cpin=1.05p rpin=1.5
lp pi mid lpin
cp mid 0 cpin
rp mid po rpin
.eom pin
*
* define parameters for VDD, package, temperature
.PARAM vdd_ref=3.15
.param vdd='vdd_ref*0.85'
.param cint=0.7p
.param lpkg=10.2n
.param cpkg=1.05p
.param rpkg=1.5
.temp 55
.param ioff = 5e-4
*
* define timing arcs
*
* measure transient arc_name trigger volt(signal) threshold
```

Example D.1 (Spice Netlist for Characterization) (Continued)

```
* direction start looking for trigger after TimeDelay end
       measurement
* at target volt(signal) threshold direction
.measure tran rise_dff trig v(q) val='.1*VDD' rise=1 TD=70
+                      targ v(q) val='.9*VDD' rise=1 TD=70
.measure tran fall_dff trig v(q) val='.9*VDD' fall=1
+                      targ v(q) val='.1*VDD' fall=1
.measure tran rise_clk trig v(clk)   val='.1*VDD' rise=1
+                      targ v(clk)   val='.9*VDD' rise=1
.measure tran Clk_QN0  trig v(clk)   val='.5*VDD' rise=1
+                      targ v(qn)    val='.5*VDD' fall=1
.measure tran Clk_QN1  trig v(clk)   val='.5*VDD' rise=1 TD=70n
+                      targ v(qn)    val='.5*VDD' rise=1 TD=70n
*
* supply instance name, pos terminal, neg terminal, DC voltage
Vdd VDD 0 VDD
* note: node 0 defined as ground
* Define stimulus vectors: instance name, pos terminal, neg terminal
*(init volts, pulse volts, delay, rise, fall, width, repeat period)
Vcl inclk 0 PULSE (VDD 0 9.9n 0.2n 0.2n 9.9n 20n)
Vin data 0 PULSE (0 VDD 0n 1n 1n 59n 120n)
Ven ena 0 PULSE (0 VDD 20n 1n 1n 19n 60n)
*
.TRAN 0.02n 120n
*
* The simple netlist, clk buffer and dff
*
X1 0 inclk clk VDD BUFF
C0 clk 0 0.1p
X10 data ena clk q qn dff cap=cint
*
* Change parameters and rerun the simulation
.alter 'Rerun using Fast Model'
.protect
.del LIB '/libraries/models.sp' "slow"
. LIB '/libraries/models.sp' "fast"
.unprotect
.param vdd='vdd_ref*1.15'
.param cint=0.4p
.param lpkg=4.4n
.param cpkg=0.4p
.param rpkg=0.7
.temp 0
*
.END
```

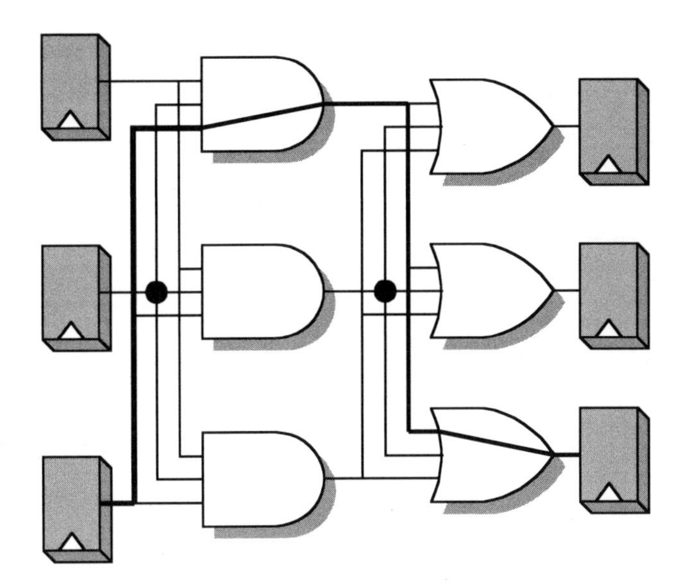

References

1. [Actel96] Actel Corporation. *FPGA Data Book and Design Guide*. Actel Corporation. 1996.
2. [Actel98] Actel Corporation. *Online Help*. Actel Corporation. 1998.
3. Altera Corporation. "FLEX 8000 Programmable Logic Device Family Data Sheet." September 1998, Version 9.11.
4. [Altera98] Altera Corporation. "Understanding FLEX 8000 Timing." Application Note 76, January 1998, Version 2.
5. Altera Corporation. "MAX+PLUS II Programmable Logic Development System, Getting Started." September 1997, Version 8.1.
6. M.G. Arnold. *Verilog Digital Computer Design—Algorithms to Hardware*. Upper Saddle River, NJ: Prentice Hall. 1999.
7. A. Aziz, J. Kukula, and T. Shiple. "Hybrid Verification Using Saturated Simulation." 35th Design Automation Conference. San Francisco, CA. 1998.
8. M. Bolton. *Digital System Design with Programmable Logic*. Reading, MA: Addison-Wesley. 1990.
9. [Chiprout98] E. Chiprout. "Interconnect and Substrate Modeling and Analysis: An Overview." *IEEE Journal of Solid-State Circuits*. Vol. 33, No. 9, September 1998.

10. [Dang81] R.L.M. Dang and N. Shigyo. "Coupling Capacitances for Two-Dimentional Wires." *IEEE Trans. Electron Devices Lett.* EDL–2:196–197, 1981.

11. F. Dartu and L.T. Pileggi. "TETA: Transistor-Level Engine for Timing Analysis." 35th Design Automation Conference. San Francisco, CA. 1998.

12. S. Devadas, A. Ghosh, and K. Keutzer. *Logic Synthesis*. McGraw-Hill, Inc. 1994.

13. Exemplar. *Leonardo Synthesis and Technology Guide*. Exemplar. 1998.

14. [GD85] L.A. Glasser and D.W. Dobberpuhl. *The Design and Analysis of VLSI Circuits*. Reading, MA: Addison-Wesley. 1985.

15. F.J. Hill and G.R. Peterson. *Digital Logic and Microprocessors*. John Wiley. 1984.

16. M. Keating and P. Bricaud. *Reuse Methodology Manual for System-on-a-Chip Designs.* Norwell, MA: Kluwer Academic Publishers. 1998.

17. [KA97], P. Kurup and T. Abbsi. *Logic Synthesis Using Synopsys*. 2d ed. Norwell, MA: Kluwer Academic Publishers. 1997.

18. [LB94] W.K.C. Lam and R.K. Brayton. *Timed Boolean Functions: A Unified Formalism for Exact Timing Analysis*. Norwell, MA: Kluwer Academic Publishers. 1994.

19. [Lewis84] E.T. Lewis. *An Analysis of Interconnect Line Capacitance and Coupling for VLSI Circuits*. Solid State Electronics, 1984.

20. M.M. Mano. *Computer System Architecture*. Englewood Cliffs, NJ: Prentice Hall. 1982.

21. M. Nemani and F. Najm. "Delay Estimation of VLSI Circuits from a High-Level View." 35th Design Automation Conference. San Francisco, CA. 1998.

22. S. Palnitkar. *Verilog HDL: A Guide to Digital Design and Synthesis*. Upper Saddle River, NJ: Prentice Hall. 1996.

23. Quad Design Technology. *Motive User's Manual*. Camavillo, CA: Quad Design Technology, Inc. 1996.

24. [SK93] S.S. Sapatnekar and S.M. Kang. *Design Automation for Timing-Driven Layout Synthesis*. Norwell, MA: Kluwer Academic Publications. 1993.

25. M. Sivaraman and A.J. Strojwas. *A Unified Approach for Timing Verification and Delay Fault Testing*. Norwell, MA: Kluwer Academic Publishers. 1998.

26. [Smith97] M.J.S. Smith. *Application-Specific Integrated Circuits*. Reading, MA: Addison-Wesley. 1997.

27. Synopsy. *Synopsy PrimeTime User Guide*. Mountain View, CA: Synopsy. 1998.

28. [TM96] D.E. Thomas and P.R. Moorby. *The Verilog Hardware Description Language*. 3d ed. Norwell, MA: Kluwer Academic Publishers. 1996.

29. [Wilnai71] A. Wilnai. "Open-Ended RC Line Model Predicts MOSFET IC Response." *EDN/EEE*, pp. 53–54, December 15, 1971.

30. [Xilinx99] Xilinx, Inc. "A Look at Minimum Delays," Application Note. Xilinx, Inc. 1999.

31. Xilinx, Inc. *The Programmable Logic Data Book*. Xilinx, Inc. 1998.

INDEX

A

B

C

D

E

F

G

H

I

L

M

N

O

P

Q

R

S

T

V

W

X

Z

About the Author

Farzad Nekoogar is Director of Design Services at Silicon Designs International. Farzad has extensive practical experience verifying timing of ASICs, FPGAs, and systems-on-a-chip. He is the author of *Digital Control Using Digital Signal Processing,* published by Prentice Hall PTR. He has lectured at the University of California at Berkeley on signal processing, control systems, and theoretical physics (specifically, Superstring Theory). He is currently a lecturer at the Department of Applied Science at the University of California at Davis.

Farzad, seen here in December 1992 at Stanford University, with Sir Roger Penrose. Farzad writes: "In this book we try to solve timing issues related to design of microchips. I am honored to be pictured here with Sir Roger Penrose, one of the most brilliant scientists of all time, who has authored some of the most complex theories about space-time, contributing a lot to our understanding of the universe."

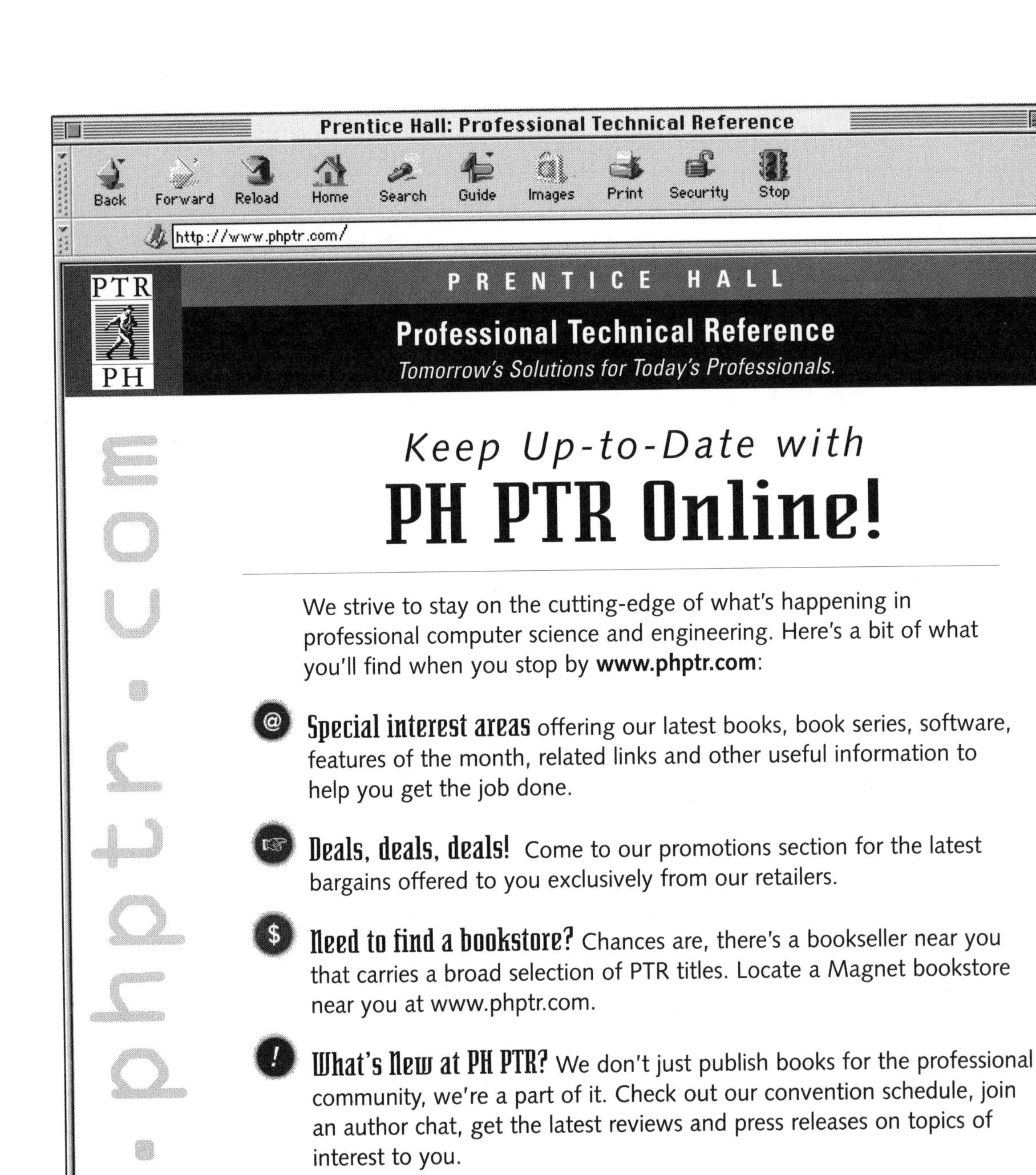
Prentice Hall: Professional Technical Reference
Back
Forward
Reload
Home
Search
Guide
Images
Print
Security
Stop
http://www.phptr.com/
PTR
PH
PRENTICE HALL
Professional Technical Reference
Tomorrow's Solutions for Today's Professionals.
www.phptr.com
Keep Up-to-Date with
PH PTR Online!
We strive to stay on the cutting-edge of what's happening in professional computer science and engineering. Here's a bit of what you'll find when you stop by www.phptr.com:
Special interest areas offering our latest books, book series, software, features of the month, related links and other useful information to help you get the job done.
Deals, deals, deals! Come to our promotions section for the latest bargains offered to you exclusively from our retailers.
Need to find a bookstore? Chances are, there's a bookseller near you that carries a broad selection of PTR titles. Locate a Magnet bookstore near you at www.phptr.com.
What's New at PH PTR? We don't just publish books for the professional community, we're a part of it. Check out our convention schedule, join an author chat, get the latest reviews and press releases on topics of interest to you.
Subscribe Today! Join PH PTR's monthly email newsletter!
Want to be kept up-to-date on your area of interest? Choose a targeted category on our website, and we'll keep you informed of the latest PH PTR products, author events, reviews and conferences in your interest area.
Visit our mailroom to subscribe today! http://www.phptr.com/mail_lists